PERSPECTIVES ON WILDLIFE SOUND RECORDING

An Overview of Modern Practice

EDITED BY MARK FERGUSON

PELAGIC PUBLISHING

First published in 2026 by
Pelagic Publishing
20–22 Wenlock Road
London N1 7GU, UK

www.pelagicpublishing.com

Perspectives on Wildlife Sound Recording: An Overview of Modern Practice

https://doi.org/10.53061/SIIB5831

A CIP record for this book is available from the British Library

ISBN 978-1-78427-621-8 Pbk
ISBN 978-1-78427-622-5 ePub
ISBN 978-1-78427-623-2 PDF

EU Authorised Representative: Easy Access System Europe –
Mustamäe tee 50, 10621 Tallinn, Estonia, gpsr.requests@easproject.com

Front cover photographs (left to right, top to bottom): Izabela Dłużyk (Izabela Dłużyk), Emmanuel Baffoe (Justice Ayertey), Jana Winderen (Stuart Trowell), Matt Mikkelsen (Palmer Morse), Jo Kennedy (C. M. Shaw), Laura Giannone (Laura Giannone), Axel Drioli (Ario Drioli).
Frontispiece: *Song Thrush with Snail Shell*, David-Lee Badger (davidleebadger.com)
Back cover photograph: Mark Ferguson (Mark Ferguson)

Index by Jeremy Complin

Typeset in ITC Stone Sans Std and DIN2014 by S4Carlisle Publishing Services, Chennai, India

5 4 3 2 1

Printed in the UK by Hobbs the Printers Ltd

MIX
Paper | Supporting responsible forestry
FSC® C020438
www.fsc.org

CARBON BALANCED PRINT
www.carbonbalancedprint.com
CBP2250

PERSPECTIVES ON WILDLIFE SOUND RECORDING

For Krysia

Download the audio library here:
https://pelagicpublishing.com/pages/perspectives-wildlife-sound-recording-resources

Contents

About the Editor

Dr Mark Ferguson has been recording wildlife sounds since 2014. His work has been featured by the BBC, mentioned by *The Guardian* and played in major arts and academic institutions worldwide, including Stanford University, the Royal College of Music in Stockholm, the University of Birmingham, and the Barber Institute of Fine Arts. He has received two scholarships from the Arts and Humanities Research Council for postgraduate studies in electroacoustic composition, and has won multiple awards and commendations for his recordings, notably in the Wildlife Sound Recording Society and Phonurgia Nova annual competitions.

As a freelancer, he has provided post-production services for established broadcasters and independent UK filmmakers; more recently, he has worked for a number of leading audio manufacturers, offering high-level practical advice and technical guidance to some of the world's top sound professionals.

He is currently based in Northern Ireland, where he spends most of his time documenting the habitats and hidden stories of his native Ulster countryside.

Perspectives on Wildlife Sound Recording is his first book.

markfergusonaudio.com

Contributors

In order of appearance:

Izabela Dłużyk

Poland

Website: zvukolom.bandcamp.com/album/soundscapes-of-summer

Izabela was born in Gdańsk, Poland. Blind since birth, she showed an interest in the natural world throughout her childhood, developing a particular fascination with avian species and their vocalisations. Guided by her unique understanding of sound, she spent years developing her craft using low-budget equipment, learning as much as possible about wildlife through listening and recording. During her university studies, she began working with professional audio equipment and has since devoted much of her time to depicting the natural sonic beauty of her native Poland, as well as recording in the tropical rainforests of Peru and Malaysia. Through the practice of sound recording, she attempts to document and share the subtleties of nature, exploring the inherent beauty of the world's habitats and wild species. (Bio photo: Izabela Dłużyk)

Simon Elliott

UK

Website: xeno-canto.org/contributor/FFFADKCCII?view=1

Simon has been recording wildlife sounds since 1968. He has recorded on six continents and has contributed sounds for a range of film, TV, radio, book, CD, internet and art installation projects. He has won a number of sound recording and documentary competitions/awards, including: the Wildlife Sound Recording Society (WSRS) Annual Competition (four times overall winner and Silver Fox trophy recipient); British and International Amateur

Recording Contests; and the GP Audio/Prix Jean Thévenot of the Fédération International des Chasseurs de Sons (2016). His recent projects include detailed studies of moorhen, jackdaw and owl vocabulary around his home in rural Yorkshire. (Bio photo: Julie Elliott)

Samuel Kudjodzi

Ghana
Website: linktr.ee/samuelkudjodzi

Samuel is an acoustic ecology researcher, field recordist and media educator. He is an alumnus of Bauhaus-Universität Weimar, where he studied Electroacoustic Composition, Sound Design and Instructional Design. His key areas of interest include: digital interactive sound; documentation, archiving and analysis of complex acoustic spaces; ecological and acousmatic listening; ethnoacoustics; experimental electroacoustic music; urban soundscape studies; and the development of curricula in the fields of acoustic education and digital interactive media. He is founder of the Centre for Soundscape Ecology and Multimedia and is a member of Quiet Parks International. (Bio photo: Bauhaus-Universität Weimar)

Jo Kennedy

UK
Website: jokennedysound.com

Jo lives in the South Pennines. She trained in environmental science, spending over two decades working in the UK environmental sector before returning to university in her late forties to study for a degree in Creative Music Technology. She now works as a freelancer, using field recordings, found sounds, spoken word and music to make immersive audio experiences. Her own creative practice engages explicitly with ecological issues, responding to questions about our relationship with landscape and the living world. She co-produces the *Nature Tripping* podcast and is also a director at Yorkshire Sound Women Network: a community interest company working on initiatives to address gender inequality in the audio industry. (Bio photo: C. M. Shaw)

Greg Green

UK

Website: soundcloud.com/greggreensound

Greg is a UK-based sound recordist and audio cataloguer. In his day job at the British Library Sound Archives, he preserves and catalogues thousands of unpublished historical wildlife recordings from around the world. He has worked on radio programmes, online exhibitions and several installations, including a multichannel adaptive sound installation for the British Library exhibition *Animals: Art, Science and Sound* (2023). When not working with other people's sounds, he enjoys recording and studying wildlife in the UK, occasionally venturing farther afield to locations such as Sweden and Mongolia. He has won multiple awards in the WSRS Annual Competition, including the Richard Margoschis Documentary Competition trophy for his short piece 'In Pursuit of the Nightingale', which was produced during the 2020 COVID-19 lockdown. (Bio photo: Alan Burbidge)

Nikki Sheth

UK

Website: nikkisheth.myportfolio.com

Nikki is an internationally recognised sound artist and composer, whose interests include interspecies communication, nocturnal soundscapes and acoustic ecology. Her work aims to give voice to the environment and foster deeper connections to the natural world through field recording, soundscape composition, spatial audio practice, multimedia installation and soundwalking. She has received multiple awards and nominations, and her work has been presented globally at venues/events such as the Quench Gallery (Margate), Kings Place (London) and Ars Electronica (Austria). Nikki's sounds have been described by *The Wire* as 'enchanting', 'beautiful', and 'peaceful, serene and utterly alluring'. She holds a PhD in Musical Composition from the University of Birmingham and was a Visiting Research Fellow at the University of the Sunshine Coast, Australia for the research project *Visualising Nocturnal Soundscapes* in 2024. (Bio photo: John Lucy)

Axel Drioli

Italy/UK
Website: soundingwild.com

Axel is a director, producer, wildlife sound recordist and spatial audio designer. Drawing on his passion for wildlife and conservation, he creates compelling audio experiences which raise awareness of the natural world. His work has been featured in the Future of StoryTelling Summit (New York), the Seattle International Film Festival, Sheffield DocFest, the United Nations campus in Bonn, and Downing Street (London). He has also been involved in productions for Google, *The Guardian* and BBC Springwatch. His recent expeditions along the East Atlantic Flyway, Wings Across Continents and Sounds Across Continents, have received international recognition for their focus on migratory avian species and local community engagement. (Bio photo: Ario Drioli)

Jana Winderen

Norway
Website: janawinderen.com

Jana is a Norway-based artist with a background in mathematics, chemistry and fish ecology. Her practice draws attention to environments and species which are difficult for humans to access and listen to, including underwater habitats, the internal soundscapes of ice, and acoustic phenomena beyond the range of human hearing. Her activities include site-specific and spatial audio installations/concerts, many of which have been exhibited and performed internationally in major institutions and public spaces. Recent work includes: *The River* at Jerwood Gallery, Natural History Museum (London); 'Absent Voices' at Haus der Kunst (Munich); *The Art of Listening: Underwater* at Lenfest Center for the Arts, Columbia University (New York); *Listening through the Dead Zones* for IHME (Helsinki); *Rising Tide* at Kunstnernes Hus (Oslo); *Listening with Carp* for Now is the Time (Wuzhen); *Spring Bloom in the Marginal Ice Zone* for Sonic Acts (Amsterdam); *DIVE* at the Park Avenue Tunnel (New York); and *Ultrafield* for MoMA (New York). In 2011, she won a Prix Ars Electronica Golden Nica for Digital Musics & Sound Art. Jana is published by Touch Music/Fairwood Music (UK) Ltd. (Bio photo: Lena Winderen)

Ángel Pérez Grandi

Spain/Italy
Website: soundarkstudios.com

Ángel is a Spanish-Italian sound recordist currently based in the UK. Working primarily in film and often collaborating with artists (as well as art institutions such as TATE), his projects explore themes of time and place, drawing inspiration from wild species and their natural habitats. He runs SoundArk: a studio embodying both his passion for exploring sound as an art form, and his drive to preserve natural and cultural events through field recording. When he isn't working in the studio, Ángel is outside, listening, observing and recording. In 2023, he was commissioned by the Royal Opera House to create a sound piece for the Festival of New Choreography. In 2024, he was awarded the Richard Margoschis Documentary Competition trophy by the WSRS, for a short piece exploring the breeding activity of Barn Swallows. (Bio photo: Ángel Pérez Grandi)

Matt Mikkelsen

USA
Website: sprucetone.com

Matt is a field recordist, sound designer and documentary filmmaker based in Duluth, Minnesota. He has worked as a sound designer, recordist, director and producer on several award-winning documentaries, and has devoted much of his professional career to observing, recording, researching and preserving natural soundscapes. His clients include Skywalker Sound, *Emergence Magazine*, Microsoft and Amazon, and his work has been featured on NPR's All Things Considered, WBUR's On Point, New York Times Op-Docs, National Geographic and CBS Mornings. Matt is the co-founder of Spruce Tone Films and serves as the executive director of Wilderness Quiet Parks for Quiet Parks International. (Bio photo: Palmer Morse)

Laura Giannone

USA
Website: ebbtidesound.com

Laura is a field recordist and sound editor currently based in the Mojave Desert. Twelve years ago, she had a casual job reviewing nature sound libraries which evolved into editing work. Curious to find out more, she enrolled in a US college which had synth labs, recording studios, lots of audio equipment (including parabolic reflectors and standalone acoustic devices) and a fantastic Ornithology and Recording Arts faculty. She worked in campus media production and learned a lot from talented staff and co-workers, spending as much time as possible using the audio facilities and borrowing field recording gear. Since then, she has undertaken a diversity of work, including acoustic data collection and sound editing. Her family, partner, friends and colleagues cheer her on and support her in all that she does. Their love is palpable, even when she is working alone in remote places. (Bio photo: Laura Giannone)

Seán Ronayne

Ireland
Website: irishwildlifesounds.com

Seán is a Cork-born ornithologist, wildlife sound recordist, award-winning author and environmental activist. He studied Zoology (BSc) and Marine Biology (MSc) at University College Cork, later gaining an additional master's degree in Environmental Impact Assessment. After securing his first professional ornithology job in 2017, he moved to Barcelona where he began recording birds and wildlife in earnest. In 2021, he returned to Ireland with the goal of documenting the vocalisations of all commonly occurring avian species in the country, amassing a collection of over 12,000 recordings and featuring in the acclaimed documentary *Birdsong* (2024). His first book, *Nature Boy* (2024), won Irish Biography of the Year and was shortlisted for Irish Book of the Year. Seán lives near Cobh, County Cork with his family. (Bio photo: Chris Maddaloni)

Johannes (Hans) van den Berg

South Africa/UK

Email: scatterlingofafrica@icloud.com

Currently based in England, Hans spent the first 32 years of his life in South Africa. Encouraged by his parents to play in (and learn about) the natural world as much as possible, he spent his childhood with his bare feet firmly in contact with the South African soil, climbing trees and exploring the country's breathtaking scenery and wildlife. Through his knowledge of the natural world and sonic/photographic documentation of his experiences in the field, he hopes to inspire admiration and appreciation for the species we share our planet with. He is happiest when out in the African bush, recording wildlife sounds and taking photos. (Bio photo: A. T. Lada)

Andy Martin

USA

Website: andymartinnaturesound.com

Andy is an award-winning nature sound recordist and sound designer whose work has featured in video game, VR, film, television, post-production, radio and arts projects. Andy dedicated his early professional career to creative sound design in film, television and interactive media. In 2015, he launched Northwest Soundscapes as part of a career-changing quest to record quiet natural sounds around his adopted Pacific Northwest homeland in the USA. Over time, he has shared his work widely and has contributed to a range of projects as an ambience designer, field recordist and sound librarian. He continues to record quiet natural spaces and wildlife around the world, using storytelling and sound to raise awareness of conservation goals and inspire a deeper appreciation of nature. (Bio photo: Andy Martin)

Helen Hedworth

UK
Website: linkedin.com/in/helenhedworth

Currently based in Wales, Helen is a biologist, acoustician, sound recordist and former officer of the WSRS. She studied zoology and marine ecology, and worked on international field research projects on Meerkats, reef fishes and bats. She then moved into ecological survey work for a range of terrestrial and freshwater species groups, to inform planning applications for development projects. After completing her offshore training, she began acoustic monitoring of marine mammals as well as measurement of underwater construction noise. Today, Helen carries out assessments of anthropogenic noise from marine infrastructure projects and its potential effects on animal populations, and visits seabird islands to record whenever she can. She is also interested in how nature connection improves human health and wellbeing, recently spending a year as a research and evaluation coordinator for a charity reconnecting people with woodlands. (Bio photo: Helen Hedworth)

Foreword

The human desire to record the sounds of the natural world has a long history. Music and the written word were initially used to achieve this, but while both methods had their merits, neither had the ability to accurately reflect the myriad of sounds produced by animals and their wider environments. The song of the Common Nightingale, for example, has for centuries been celebrated for its richness of voice and complexity of content. Many have tried to replicate its beauty using instruments or prose; however, there is something special about hearing the sound in its original form, and the same is true regardless of whether listening to a songbird or a water beetle. It wasn't until the latter half of the nineteenth century that the ability to record and then play back the sounds of wildlife became a reality. The technology which enabled this had, for the most part, the ability to create a replica of the real sound, preserved in time for others to listen to, learn from and enjoy.

The first known sound recording of an animal (a short snippet of captive White-rumped Shama song) was made in Germany in 1889, by an eight-year-old boy named Ludwig Koch, using a phonograph and wax cylinder given to him by his father. This gift would shape Koch's entire life, generating a deep and enduring passion for sound that would see him become one of the great pioneers in the field of wildlife sound recording. And Koch was not alone: during the first decades of the twentieth century, he and other figures across Europe and North America began to explore the possibilities of recording wild animals in their natural habitats. Though usually working independently from each other, these pioneers slowly but surely started to overcome the technological, logistical and biological challenges encountered when trying to record wildlife in the field, paving the way for those who would follow.

As with many other activities that seek to document moments in time, individuals who record wildlife sounds are usually absent from their recordings. These silent collectors unselfishly remove themselves from what has been documented, wanting the focus to be entirely on the featured species or habitat. Sometimes their names are known, especially if the recordings are published or archived with a public institution; however, there is rarely the opportunity to learn more about the recordists themselves, especially in their own words.

The 15 practitioners featured in this collection of interviews present a rich and varied account of wildlife sound recording today. From working in challenging environments and setting up equipment to connecting with people and finding peace within nature, this diverse group of individuals dedicate so much of their time to documenting and understanding the role sound plays in the lives of animals. Some are hobbyists, while others are artists; some are fascinated by vocalisations that are difficult to hear, while others are drawn to those wider sounds where a range of voices come together to create a soundscape. Common themes emerge as we read, and it becomes apparent that these recordists share a number of characteristics, including patience, ingenuity, adaptability and a deep love of the natural world. The interviews also show that wildlife sound recording is an activity for everyone, regardless of background, gender, ethnicity or physical ability. Though challenges remain, it is encouraging to see the field becoming a more inclusive space.

As wildlife sound recording enters its second century, *Perspectives* encourages us to take a moment to reflect on how far the craft has come and to consider where it might be going. Thanks to the dedication and skill of today's practitioners, the future looks full of opportunities that will undoubtedly take the field further than ever before.

Cheryl Tipp
Wildlife and Environmental Sounds Curator
The British Library

Acknowledgements

First and foremost, I would like to thank my incredibly supportive family and friends: without your love, respect, words of wisdom and banter, I simply wouldn't have been able to do this.

Next, I would like to offer my heartfelt thanks to Nigel Massen and everyone at Pelagic for their unwavering support, from start to finish. Dozens of questions were patiently and professionally addressed over the course of editing, and all of my ideas were listened to with attentive ears. I genuinely can't imagine a better publishing team to work with.

It's impossible to express how grateful I am to each and every recordist who participated in this book. Izabela, Simon, Samuel, Jo, Greg, Nikki, Axel, Jana, Ángel, Matt, Laura, Seán, Hans, Andy and Helen: it has been an absolute pleasure to share your thoughts and stories. Thank you for your remarkable patience, understanding and hard work throughout.

I would also like to thank, in no particular order:

Clive Davies and Alan Burbidge, for listening to my earliest ideas for *Perspectives* and encouraging me to develop the project in the most suitable way; Cheryl Tipp, for writing the foreword and assisting with archived recordings from the British Library; Andy Martin, Simon Elliott, Ángel Pérez Grandi and Prof Scott Wilson, for having a look at the glossary and making helpful suggestions; Dr Sue Gould, Prof Annie Mahtani and Thor Hanson, for reading through everything in detail and offering words of encouragement; Ian Brady, for his understanding and help in connecting me with the right people; all members of the WSRS; and David-Lee Badger, for the superb Song Thrush frontispiece.

Curated texts like *Perspectives* are inherently selective, and choosing fellow recordists for inclusion in the project has genuinely been one of the toughest aspects of editing; ultimately, it wasn't possible to involve as many people as I would have liked. To anyone who was contacted in the preparatory stages of the book but, for one reason or another, couldn't be included: thanks for your time and understanding.

Finally, I would like to thank you, the reader-listener. Perhaps you are already familiar with wildlife sound recording; perhaps you are just starting out on your journey. Whatever the case, I hope you enjoy this book.

Mark Ferguson
Belfast, September 2025

PART I

INTRODUCTION

Preamble

About the Text

This book aims to provide a representative overview of modern wildlife sound recording practice, through a series of one-to-one interviews with leading recordists and audio professionals around the world. Accompanied by 121 photos,159 high-quality audio tracks and a concise glossary of terms and definitions, it explores some of the key ideas and methodologies underpinning our craft, offering an informative cross section of the wildlife sound recording community today.

Topics covered include:

- Audio techniques/technologies and their practical implementation through fieldcraft.
- Trip planning, logistics and personal safety.
- The relationship between wildlife sound recording and related areas such as bioacoustics, audio post-production and nature conservation.
- General perspectives on craft, including practitioner motivations, inspirations, and thoughts on listening to the natural world.

Motivations and Objectives

My primary motivator for putting together *Perspectives* was the realisation that a detailed reference text focusing on wildlife sound recording practice from a community angle has never been published. There are, of course, some very helpful and informative social media groups, blogs, mailing lists, members' forums and other online resources. A few good books exist, but many are dated and most have been written by individual male practitioners. With the exception of one or two interview-based titles covering the wider approaches of field recording, no single book provides an open platform for leading wildlife sound recordists to discuss their methodologies in detail and convey the personal motivations bound up with their work.

My approach to *Perspectives* has been underpinned by three fundamental objectives, all of which have guided my editing and shaped the overall trajectory of the finished book:

Firstly, I wanted to curate material that was diverse and inclusive, incorporating viewpoints from people of various cultural backgrounds, ages, gender identities, physical capabilities and ways of thinking. In parallel, I wanted to expand the published discourse on wildlife sound recording so that the contemporary scope of the craft (which now encompasses natural-history production work, bioacoustics, sound art, music composition, video game development, nature conservation, community-based outreach programmes and a host of other dynamic areas) could be more fully appreciated.

Secondly and with my recordist hat on, so to speak, I wanted to create some space for interviewees to discuss the personal motivations which feed into their practice: childhood influences, outreach projects, involvement in charitable organisations, awareness-raising activities and other elements that matter deeply to so many members of our community. The craft of wildlife sound recording does not exist in a vacuum; in order to understand it as fully as possible, we need to listen to the people who practice it, moving beyond the exploration of techniques and technologies towards human context. This can only happen if people are given the freedom to articulate what motivates and inspires them, and share thoughts on pertinent issues within the field recording milieu and the wider audio industry.

Finally, I wanted to establish a transparent interview framework where each and every recordist could speak openly about their work, without any form of judgement or criticism. The views expressed (and methodologies discussed) by the people featured in this text are thus very much their own, and they emerged naturally from the questions I developed around known areas of interest and expertise. Regarding my own interpretation of interview responses, I refrained from drawing conclusions about what was 'right' or 'wrong' in wildlife sound recording. I also decided at a very early stage not to write an editor's conclusion for the text: the interviews do this job better than I ever could, serving as a nuanced, up-to-date snapshot of our practice and the fascinating and varied processes within it.

Who this Book is for

Perspectives will hopefully make an informative and insightful reference text for anyone interested in recording the sounds of wild species and their habitats.

It is particularly suitable for:

- Wildlife sound recordists of all ages and abilities, seeking to explore fellow practitioners' methodologies and viewpoints on craft.
- Those involved in related areas of field recording who would like to understand more about the specialised techniques underpinning wildlife sound recording practice.

- Ecologists, bioacousticians and other scientists whose work incorporates sonic identification or monitoring of wild animals (e.g. acoustic surveys, behavioural studies of species).
- Sound designers and other post-production experts interested in recording wildlife, or curious about the field approaches underpinning the species- and habitat-based assets used for their own projects.
- Educators looking for an inspiring and instructional text focusing on wildlife sound and the people involved in it.

On a much broader note, the book aims to inspire general curiosity about our sonic environment and the audible behaviours of undomesticated species. In this sense, it is suitable for absolutely anyone (regardless of age, background or technical ability) with an interest in listening to the natural world.

How to Use the Book

Since *Perspectives* offers a community-focused cross-section of modern wildlife sound recording practice, it is not intended to be read from start to finish: interviews can be freely dipped into based on the interests of the reader.

Basic familiarity with audio technologies and field recording techniques (e.g. microphone types, recording configurations, portable recorder operation) will be useful, but is not assumed. For those unfamiliar with these areas or anyone who simply wishes to refresh their audio knowledge, a concise glossary has been provided at the end of the main text.

Editorial Approaches

Interview Methodology and Ethical Considerations

Any book based on other people's work can only proceed on the basis of mutual trust and respect; with this in mind, all editing was undertaken with as much transparency as possible.

After deciding on a list of contributors, I contacted everyone directly via email and/or video call to explain the nature of the project and my main objectives with it. Points of focus for the interview were discussed and agreed upon, and questions were drafted around these. A full list of contributors was shared, and it was made clear that each recordist would be giving their time independently and voluntarily. The interviews were then conducted over a period of weeks via email and video call, in various combinations.

Edited interviews, bios, photo captions, track notes and audio files were sent to each contributor for approval, and each person was offered the opportunity to make revisions or changes. Very minor spelling, grammatical and structural revisions were then made on my side, before the manuscript was prepared for final publication by Pelagic.

To minimise my overall carbon footprint and environmental impact throughout the project, all editing work was undertaken from my home studio in the UK, using standard word-processing software without any form of AI assistance.

Brand and Product Generalisations

Throughout *Perspectives*, all mentions of audio equipment (microphones, portable recorders, windshields, field accessories, etc.) have been edited to reflect the general technology or equipment type used, rather than branded product(s).

My intentions behind this were twofold: to make the book as accessible as possible, especially for those unfamiliar with audio technologies; and to preserve meaning within technically orientated portions of the interviews well into the future (products inevitably have a lifespan, and some may no longer be relevant or known about years down the line). By way of example: 'I used a cardioid pencil-condenser microphone placed inside a tripod-mounted basket windshield' arguably carries much more meaning than simply stating a string of product names; in addition, this sentence will probably still make sense well beyond the book's publication date, even if the manufacturer products associated with it change.

Whilst I appreciate this 'brand agnostic' approach may be disappointing to some readers (especially those interested in sound recording technologies), in this book, I wanted to keep the focus firmly on wildlife sound recording *practice*: how technologies are actually selected and applied, fieldcraft considerations, and the community perspectives bound up with our work. I am certainly not alone in feeling that these elements are frequently lost in extensive and highly subjective conversations (notably on online forums, social media groups and blogs) about which branded products were used to achieve a particular result.

I should also mention that just because product names have been removed from the main text doesn't mean that there is anything wrong with product-driven discussions; indeed, I enjoy conversations about branded audio gear as much as the next recordist. As a concluding point, it is perhaps enough to say that branded products were simply never intended to be a primary focus of *Perspectives*. Instead, the generalised descriptions of audio technologies and methods within the text represent an accessible starting point for enquiries into branded solutions via manufacturer knowledge bases, support channels, and the wider field recording community.

Geographical Scope and Representation

Perspectives features recordists from Ghana, Ireland, Italy, Norway, Poland, South Africa, Spain, the UK and the USA, all of whom are actively listening to and recording wildlife in various locations around the globe. Recent and ongoing work across six continents is discussed, incorporating insights from the Mongolian steppes, the Arctic, European wetlands, African deserts, some of the United States' most famous national parks, and many other locations and habitats. With this in mind and to help convey the geographical scope and diversity of the text, all interviewees were asked to specify a country (or countries) which they felt best represented their national identity, and/or their association with current working areas. These country names are included next to each person's bio in the list of contributors section.

Species names have been edited to broadly correspond to the classification principles of the IOC World Bird List. For example, 'Robin' (*Erithacus rubecula*) has been expanded to its full common English name of 'European Robin'. Whilst this may occasionally come across as unnatural in the general flow of conversation, the intention behind this approach has been to make all species references clear and understandable in an international context.

About the Accompanying Resources

Downloadable Audio Library

The book is accompanied by a downloadable library of 159 instructional wildlife recordings and other audio excerpts, most of which are cited in the main flow of the text. A PDF document containing supplementary track notes has also been included in each interviewee's audio folder.

The following format has been adopted for filenames in the audio library (modified slightly for creative works, podcasts, etc.):

```
Recording Number - Subject Name (Optional Extra Info) -
Recordist Name - Location
```

For example, this audio citation in Izabela Dłużyk's interview

> I remember recording a European Robin, which sat on my microphone and sang several phrases of very quiet subsong (Audio 1.6).

would correspond to the following recording filename in her audio folder:

```
1.6 - European Robin - Izabela Dłużyk - Poland.wav
```

With regard to editing audio materials submitted by interviewees, my guiding principle was to leave files in their original state, wherever possible. LUFS-based loudness mastering was applied as needed, in the form of overall level adjustment and nothing more. Terminal fades were added at the beginning and end of a handful of files, and one or two very long recordings were excerpted. Apart from these minor changes, no other adjustments were made.

All tracks in the library are suitable for playback using consumer-quality headphones and home loudspeakers; however, to fully appreciate the intricate details of the wildlife behaviours featured, high-quality headphones or studio monitors are strongly recommended. Regardless of the quality of listening equipment used, a quiet environment is essential to absorb the fascinating and highly detailed sounds featured throughout the text.

Please follow this link to access the audio library: https://pelagicpublishing.com/pages/perspectives-wildlife-sound-recording-resources

Glossary

Explanations in the glossary have been limited to key terms and definitions appearing in the main text and supplementary track notes, with a view to providing a very basic reference point for readers unfamiliar with wildlife sound recording practice. Further information on all glossary terms is readily available online, although at the time of writing, caution is urged when using search engines and AI-driven summaries/responses to guide learning.

The subscription-based WSRS represents an excellent complementary resource for learning about the terms covered in the book, and many society members have decades of field experience to share alongside explanations of techniques and technologies pertinent to wildlife sound recording. For more information on how to join, visit wildlife-sound.org.

PART II

INTERVIEWS

Izabela Dłużyk

Izabela Dłużyk is a recordist with a deeply personal connection to sound. Fascinated by wildlife since childhood, for her, exploring nature is about 'enjoying the smallest things and contemplating beauty, experience and feeling'. Her appreciation and respect for the natural world are underpinning themes throughout this interview, which explores (amongst other topics) her early sonic influences, listening approaches in the field, memorable moments with nightingales, and recording activities throughout Poland. Having listened to many of Izabela's sounds over the years and spent some time recording wildlife myself in Poland, I was especially interested to read about her fieldcraft approaches for locations such as the Biebrza Marshes and Białowieża Forest, as well as her experiences with notable central/eastern European species such as European Bison and Thrush Nightingale. *MF*

How did you become interested in wildlife sound recording? What inspires you and why do you do what you do?

My early inspiration came when I was about ten years old, from some low-quality birdsong recordings I heard whilst browsing through an encyclopaedia on a CD-ROM. I started memorising each song and soon wanted to find more, so I ended up buying a few cassette tapes featuring Polish bird vocalisations. At around the same time, I was gifted an additional tape containing amphibian sounds, and those fascinated me as well.

Eventually, I decided to explore the sounds I could hear outside, and that's how I made my first field recordings. I had a very noisy tape recorder, so I considered it a success if I could hear a species clearly when the sounds were played back! The limitations of my equipment meant that I had to approach birds as closely as possible in the field, because each subject had to be near the microphone to be audible. In the beginning, I recorded everything I could. I remember trying to record sounds from a flock of House Sparrows high up in a tree, by climbing on top of a fence to get a little bit closer and holding a branch for balance. There is nothing like childhood adventure...

Nature inspires me in the sense that listening to it—and depicting its sounds through recording—is like delving into a mysterious new world. Exploring nature is about enjoying the smallest things and contemplating beauty, experience and feeling. It is about expanding our hearts, and I believe we will remain human as long as we cherish our ability to become amazed and moved by the natural world.

What are your main approaches as a recordist? Do you have any personal listening methods or areas of interest?

I'm interested in discovering and sharing glimpses of animal behaviour. This requires a certain amount of knowledge about species and their habitats, which I am constantly trying to broaden. When I can, I try to document activities such as feeding, bathing and offspring-rearing, being very careful not to disturb the species I approach.

When it comes to choosing a particular spot for recording, I often rely on spontaneous impressions: the extraordinary richness of a dawn chorus, an especially pleasing set of songbirds, an impressive gathering of a particular species (Audio 1.1), or other acoustic phenomena. There might be a gentle breeze, trickling water or an interesting echo that catches my interest and leads to further exploration. One thing I can say for certain is that I am not interested in collecting huge numbers of recordings, or accumulating as many rare wildlife sounds as possible. I'm definitely not interested in ticking species off a list. If I have an opportunity to record a rare species, I'll sometimes take it, but this is definitely not my main goal; instead, my main wish is that my recordings have something captivating to offer to the listener, something to admire and reflect upon.

In terms of recording length, I feel that nature is not to be rushed. I prefer making recordings of at least five minutes' duration, where possible. Careful listening requires patience: unlike photography, you can't just skip through a series of recordings and grasp their meaning in a few seconds. Personally, I feel that it is very important to take time to be immersed in sound, to receive and digest it. Sound is dynamic and changes over time, weaving a narrative (as with a book or film). This is something that deserves to be explored.

Izabela, holding a parabolic reflector by her side in Malaysia. (Izabela Dłużyk)

When I'm in the field, I often marvel at the musical qualities of nature. It's not just birds that interest me: recently, I had a chance to spend several days in the rainforests of Malaysia, where I recorded the evocative songs of female Agile Gibbons (Audio 1.2). Gibbon songs vary from species to species, but all of them are breathtaking and there are some studies that link these calls to the development of music. I have also recorded a Eurasian Beaver in Warta Mouth National Park (Poland), tapping its tail against the water and gnawing on logs (Audio 1.3).

I use several recording methods, so I'm not really limited by a single approach. I typically record individual species by hand, using a parabolic reflector. When working in wider habitats, I leave microphones unattended in selected spots for long periods of time, sometimes several days; this approach allows me to record sounds without creating disturbance through my presence. It's all about depicting nature from various perspectives, and I explore both of these methods based on how I feel about each moment.

I'm very interested in how you listen to and record habitats and natural phenomena, especially contextualising sounds produced by wind, flowing water and so on. Can you tell me more about how you approach these elements?

When I record species, I tend to be as flexible and open as possible when it comes to accompanying sounds, including elements such as wind and other flora/fauna in the background. I'm not interested in assembling a database of perfectly 'clean' individual-species sounds for bioacoustics purposes, so I do not need to be especially choosy when it comes to recording conditions.

Izabela takes a moment to listen to wildlife in a grassy clearing. This photo was taken in Poland's Białowieża Forest, where she has recorded European Bison, Thrush Nightingales and other species. (Izabela Dłużyk)

As I have touched on previously, I tend to accept what nature is offering me at a given moment. You cannot make nature act and behave as you wish, and this teaches us humility, which I believe is a very important trait for a wildlife sound recordist to develop. I know of some people who consider almost any kind of wind noise distracting: in fact, I once read an opinion that the rustling sound of European Aspen leaves moving in the wind often ruins recordings. This surprised me, because I regard it as one of the most beautiful plant sounds (although I haven't yet managed to make a good recording of aspen leaves).

In general, I feel that wind adds context, as long as it is not too violent or aggressive. The same goes for things like rain and snow. The only sounds I try to avoid wherever possible are human-made or human-influenced sounds, such as traffic noise or rain falling on the roof of a house. I feel that we are inundated with human sounds on a daily basis and constantly surrounded by commotion. I want to get away from this as much as possible. My philosophy is to focus on sounds in environments which are relatively free from anthropogenic activity and to portray them in their naturally occurring state; for this reason, I am not very interested in recording (for example) avian soundscapes in cities or suburban areas.

I'm curious about how other senses such as smell and touch might inform your recording practice. Can you describe how you work with these senses in the field?

As surprising as it may seem, tactility does not play any role in my recording approach, since I do not blend it with the auditory when recording in the field. I tend to focus exclusively on listening: any other sensory elements which appear are enriching, but in their own way and on a different level, which I would label as poetic and emotional (i.e. they add to my general contemplative experience).

I do not use my sense of smell for analytical purposes, although it is possible, of course. For me, smell (similarly to, or perhaps even more than, touch) is about mindfulness and cherishing the moment. I remember recording some Field Crickets in a forest clearing in the Biebrza Marshes of north-east Poland several years ago, when suddenly, I caught the scent of lilac in the wind. It must have been growing nearby. There was something mystical about that moment which moved me deeply.

The problem is that humans are unable to reproduce touch or smell in the same way that we can render sight and sound via photos and recordings. Perhaps this is why I find such tactile/olfactory experiences problematic when I try to combine them with recording experiences: they are simply too personal to be shared, and there is no reliable way to do so. Still, I sometimes regret that I do not know the names of many of the plants I can smell when recording in the field; if I did, these sensations would at least be partially describable in written or spoken form.

Pathways and boardwalks through nature reserves can offer opportunities to cover extensive portions of surrounding habitat quickly, especially when using a parabolic reflector. (Izabela Dłużyk)

Many recordists customise their audio equipment based on their approaches and target subjects. Have you made modifications to your parabolic reflectors, portable recorders or other gear?

As I am totally blind, my favourite devices are those with integrated voice guidance. This feature is not essential and I have equipment which lacks it, but it is incredibly convenient. Several years ago, I used a portable recorder which had very confusing gain knobs; in order to mark their correct positioning, I had to use small heart-shaped stickers. I should also mention my 'extremely professional' hairbands, which I use to secure the cables connecting my microphone to the recorder!

Other than these things, I haven't had to resort to any modifications.

I'd like to focus on your experiences around the Białowieża Forest. What is unique about this area and how do you record there?

My strategy in Białowieża on a given day depends on whether I want to record soundscapes or individual species. For soundscapes, I'll either explore an area on foot with a handheld parabolic reflector or leave tripod-mounted microphones unattended for longer periods, mostly from dusk to dawn. For individual species, I tend to talk to local people and ask them about desirable locations for recording (they might know of particular song-posts or display areas, for example).

In Białowieża, it's very important to act sensibly around European Bison (Audio 1.4). You really don't want to come too close to one or take a herd of them by surprise, since you might be in big trouble. People have lost their lives in such encounters. As long as you are cautious and use common sense, there is usually nothing to worry about. I once recorded a bison grazing just four metres away from me, which required constant monitoring of its body movements. I also had a situation where I could not retrieve my microphone from a dawn recording session, because there were bison grazing next to it. I managed to recover everything after the herd had moved away.

You recently made some recordings in the Peruvian Amazon and released an album of sounds from the Tambopata National Reserve. How was recording there different from your work in Poland, in terms of methods and fieldcraft?

To be honest, my approaches did not differ too much, except when it came to protecting my equipment against humidity and rain. As well as new species such as Mealy Parrots to record and learn about (Audio 1.5), there were some new safety precautions: for example, I was advised to be very careful and move slowly when near a herd of White-lipped Peccaries, since they can attack people when surprised.

Whilst in the Peruvian Amazon, I was unfortunate enough to be stung by a Bullet Ant, which is rated as having one of the most painful insect stings in the

Izabela out recording with a handheld parabolic reflector (stereo configuration) in Peru's Tambopata National Reserve. The Peruvian Amazon presented a number of fresh challenges for her in terms of equipment protection and personal safety. (Izabela Dłużyk)

world. Despite all of this, I do not regard the Amazon jungle as being particularly dangerous. I was not afraid when I went, and I would not be now.

You sometimes like to record sounds close-up. Does working in this way help you to form a sonic visualisation of the morphology [form and structure] of the species you encounter?

It is not so much a question of sonic visualisation (with the exception of wingbeats, perhaps, which give some idea of species size); instead, working close-up comes from my simple fascination with the tiniest details of sound. I remember recording a European Robin, which sat on my microphone and sang several phrases of very quiet subsong (Audio 1.6). During this encounter at a forest in Słupia, Poland, I had a wonderful sense of the bird's sounds pouring straight out of its throat and beak. The level of detail was truly astonishing.

Some people are of the opinion that recording individual species is somewhat boring and a 'well-trodden' path, especially with all of the current technical possibilities for multichannel recording. Personally, I feel that the solo song of a skylark, blackbird or any other bird is too beautiful to ever become boring, and recordings of individual species continue to fascinate me. So yes, I enjoy recording individual birds and other animals close-up, just for the sake of listening to their voices in full splendour and documenting their tiniest characteristics.

Have you had any standout species experiences as a wildlife sound recordist? Anything truly spectacular or memorable?

After many years of recording, my encounter with a Thrush Nightingale in the Białowieża Forest remains my favourite moment. I recorded it in 2015, after a year of practising with my newly bought microphone in familiar locations around my hometown. (During that year of trial and error, I learned that you cannot rely on your own hearing when using a parabolic reflector. You need to use headphones!)

Thrush Nightingales are my favourite songsters and I had always wanted to record one, but it wasn't until I visited Białowieża that I finally managed to achieve this goal (Audio 1.7). Not only was it my first time hearing a Thrush Nightingale in the field: it was also my first ever visit to the area. I hadn't been to Białowieża before, because with my previous low-quality equipment, I had assumed that the 500-kilometre journey from Gdańsk would not have been worth the hassle.

It was May, at the very height of spring. I was standing on an observation tower next to the Narewka River, listening through my headphones to the voice of the nightingale beautifully enhanced by the parabolic dish. The weather was calm, the trees were humming in the gentle breeze around me, and the air smelt heavy with dew and aquatic plants. I had been waiting for such a chance for so long. When I pressed the record button, my first thought was, 'What if my recorder breaks down just now, depriving me of this special moment?' Luckily, nothing undesirable happened and the bird's voice was preserved forever.

I also had a more recent experience involving Thrush Nightingales, which is worth mentioning. In May 2023, I visited the forests around Knyszyn, in eastern Poland. At one spot in a grassy meadow with some shrubs, I heard three male thrushes singing competitively and was struck by the amazing reverberation of their calls through a nearby forest (Audio 1.8). At first, I tried to record them with my parabolic dish on-axis, but that technique removed their voices from the beautiful spatial context. Then I decided to point the dish towards the origin of the reverberations, and the results were fascinating.

Do you have any points of advice for people interested in wildlife sound recording, based on your own challenges and experiences?

Nowadays, choosing equipment is not that important. At the moment, even an average smartphone can make acceptable recordings. It really depends on whether you want to truly devote yourself to recording wildlife or just record something pleasant from time to time.

If you are certain that you want to devote your time and effort to recording wild species and their habitats seriously, there is only one thing which is really important: you need to love it. And I mean love it, from the bottom of your heart. Only then will you learn to listen deeply, to be mindful and rediscover that often-forgotten childhood experience of feeling awe at things. Technicalities will come later. I would also say that when you aspire to do things at a professional level,

Izabela listens in a forest clearing in Białowieża. (Izabela Dłużyk)

you need to record wildlife for the sake of recording it; if, at any time, you start to expect financial gratification at every turn, you are lost. If money comes from your work, it's a bonus, but it should never be a goal in itself.

The most important element is mindset: to seek and enjoy beauty. The ability to cherish beauty is extremely important in today's world, because I think we desperately need what might be called 'mental hygiene', not only for the sake of our own wellbeing, but for the sake of those around us. If we allow ourselves to be bombarded with tragedies and crimes—to become engulfed in wars of words on social media and to constantly complain about how senseless the worldviews of our perceived opponents are—we are bound to spread negativity and bitterness to others, including our nearest and dearest. I believe we must break this cycle, and listening to nature is one of the ways that will help us to do so.

Simon Elliott

A retired consultant radiologist and long-time member of the WSRS, Simon Elliott has been recording wildlife sounds for over five decades. He has devoted much of his time to exploring the close-up details of avian vocabulary, including difficult-to-record vocalisations of birds of prey such as Osprey, Peregrine Falcon and Golden Eagle. Drawing on fieldcraft techniques developed by pioneering wildlife sound recordist Ludwig Koch, he enjoys the challenges involved in placing mics just a few centimetres away from species to document the intricacies of their behaviour. In this interview, he discusses close-miking, considerations for recording birds of prey, hide construction and unattended field techniques. He concludes with some advice for beginner wildlife sound recordists. *MF*

How did you become interested in wildlife sound recording? What inspires you and why do you do what you do?

Inspired by a BBC record of birdsong by Ludwig Koch, I made my first wildlife sound recording in 1968, aged 13, on an early portable cassette recorder. I grew up in Sheffield on the edge of the Peak District, which offered easy access to extensive

Lavalier microphones allow the recordist to get very close to target subjects without disturbance. Here, Simon places a mini omnidirectional mic at a Common Linnet song-post before cabling back to an observation site. (Simon Elliott)

woodlands and moors: the perfect playground for a young naturalist! I joined the WSRS in 1975, later serving as both journal and sound magazine editor and taking an active role in society events to the present day.

My main interest is in close-up vocabulary recording of wildlife (mainly birds), with detailed behavioural observation forming the basis for my recordings and their accompanying field notes. I love the simple approach of using distantly placed microphones and running long stretches of cable back to my observation position (Audio 2.1); this was inspired by the original techniques of Ludwig Koch, and also WSRS member Ray Goodwin, who taught me to put mics in unusual and frequently overlooked spots in the field.

I don't regard myself as a sound artist in any way. My aim is simply to record and document events in the natural world as accurately as possible, with an emphasis on quality and integrity in all of my work.

You have a particular interest in close-miking wildlife. Can you explain this approach and the fieldcraft techniques involved?

I've always loved getting as close as possible to wildlife, and I really enjoy the thrill of listening to other species without disturbing them or making them aware of my presence. I enjoy recording habitats, but for me, this never offers the same challenge or satisfaction as good-quality recordings of individual vocalisations. Wildlife sound recording often sheds light on previously undocumented sounds and behaviours, and I feel that setting out to isolate and document the complete vocabulary of each species offers tremendous potential for bioacoustics studies, helping to advance our understanding of the natural world significantly. Recording clean species vocalisations with little or no background noise (using close-miking and related techniques) is also very useful for natural history documentaries, where proximate sounds are often required for intimate wildlife scenes on-screen.

My intention is essentially to document wildlife conversations and their behavioural nuances. Imagine two ducks out on a lake: what we hear as human observers on the shore will certainly not be the same as what each bird is hearing. At the nest, calls between the adult ducks and young, including calls from within the

A miniature microphone, attached 20 cm above a Little Grebe courtship platform 50 m out from a lake shore. This placement required a waist-deep walk using chest waders, as well as cable elevation above the water using a series of embedded poles. Observation was then carried out from a natural hide without disturbing the birds. See Audio 2.2 for the corresponding recording. (Simon Elliott)

egg, would be incredibly delicate. Sounds of this intimacy have unique qualities and timbres, often much richer than when recorded at a distance; my aim in such cases would be to place my mics as close as possible to the ears of the 'intended recipient', in order to discover what they are hearing.

If you want to record these kinds of detailed sounds and behaviours, it goes without saying that good fieldcraft is essential. This comes with experience as well as accumulated knowledge on how to approach species. I always apply a simple three-step method to each field situation: observe, plan, execute. The first two (observing and planning) can take anything from a few minutes to several years. Sometimes, all three steps can take just a few minutes! For example, I remember driving along a forest trail when I saw a male Whinchat singing on top of a Hawthorn bush; after just a few moments of observation, I placed a lavalier mic about 20 cm from its perch, ran 50 m of cable back to my car and recorded for over an hour (Audio 2.3). On another occasion, I noticed a Sedge Warbler singing in a rather featureless reedbed. I quickly picked a bulrush stem, attached a mic to the top of it and stuck the base of the stem into the ground. The bird started using the improvised song-post immediately, singing all morning with its beak just a few centimetres from the mic (Audio 2.4).

My most important goal in all cases is to minimise disturbance. Ideally, the subject's behaviour should not be influenced by my own; I proceed with respect for each and every animal I record, and I regard it as a real mark of success if I don't hear any alarm calls from species I'm trying to document in the field. I'll often start by watching a singing bird to identify its favourite song-posts, as well as any areas it may be feeding, sleeping, breeding or displaying around. Then I'll try to map out a way to position a mic nearby and run a long cable back to my recorder, thinking about things like camouflage, whether other animals (e.g. grazing sheep) will get in the way of the cable or chew through it, how long I need to leave everything outside for, and so on.

In terms of equipment selection and use, when close-miking I prefer to work in mono and I'm possibly one of the few people still doing so. It certainly makes everything simpler and cheaper. If you're working really close-up, stereo can actually be a problem since the subject often ends up hopping distractingly between channels; having said this, flocks of birds, herds of deer and so on can still benefit from close-up stereo miking to convey their natural movement. Omnidirectional mics are often the best and most natural-sounding choice, and they have the added advantage of being less susceptible to low-frequency cable-borne vibrations and wind noise. Like many other recordists, I frequently use miniature omnidirectional mics intended for broadcast and film productions, and a number of companies now manufacture some very robust options at low cost. Shotgun microphones placed a little farther away from subjects also work well, and I've even left dynamic mics outside for extended periods. Ideally, I like to position a mic about 10 to 20 cm from the target species, often just above and to the side to avoid direct body contact (although it can be very rewarding to record things like a raptor's heartbeat or a displaying Western Capercaillie's feathers when a subject brushes against a mic).

I like to make my own low-cost windshields using wire or plastic garden mesh and offcuts of faux fur. I've found that birds don't like small black furry things: they'll become wary and frequently attack them, so it's good to experiment a bit with windshield colour and shape to see what blends in with the surrounding habitat. Once the mic is safely inside a windshield, I'll place it reasonably close to where I want it, then incrementally move it closer (perhaps over several days, uncoiling cable as needed) so that wildlife subjects become gradually accustomed to its shape. Good-quality cable is very important for these distant placements and I've got a total of about 1 km in storage, in varying lengths from 25 to 120 m. I don't use cable drums because of their weight and inconvenience and I always carry 25 m of cable and a small mic in my backpack, just in case I come across a close-up recording opportunity.

One thing I've learnt over the years is that any in-field connections should be kept off the ground, to avoid moisture ingress. I protect connectors with a small polythene tent and a couple of clothes pegs, which allows rain to run off and lets air circulate freely underneath. A little spray of hydrophobic lubricant around connectors can also be helpful.

This rig was used to record ducks and Whooper Swans on New Year's Day 2016, when there was minimal anthropogenic noise. The newly formed ice prevented smaller species from venturing close. (Simon Elliott)

Recording wading birds along seashores is very challenging: disturbance, overcoming noise from breaking waves and equipment risk are all factors to consider. This simple AB stereo rig (made from rust-proof angling equipment) was placed at low tide; as the tide came in, a Eurasian Curlew and other wading species moved closer. (Simon Elliott)

You've spent a lot of time over the years recording in Northumberland and have undertaken several projects focusing on raptors there. Can you tell me about the fieldcraft approaches and logistical considerations associated with recording birds of prey?

I've been lucky to spend many years recording in Northumberland and Scotland (particularly in forests, moors and on the coast), and this has naturally led to an ongoing interest in documenting raptor vocabulary: the voices of eagles, hawks, falcons and owls. The prospect of spending close-up time with birds of prey is fundamental to my enjoyment of wildlife sound recording, and it can be very challenging work. To pursue my interest in raptors, I've had to develop close relationships with field workers, scientists and organisations such as the UK Forestry Commission, as well as several private landowners. Many raptors are covered by special UK legislation and are listed as Schedule One species; to work with these at or near the nest, you require a disturbance licence from the relevant government agency (e.g. Natural England or NatureScot), for which you have to prove the necessity/importance of your project, state your experience and provide named referees.

This process doesn't come together easily and the key element underpinning all of it is trust. Most people understand what is involved in photography but few know about the requirements of sound recording, so I explain that it's in my best interest not to cause disturbance and I want the subjects to behave naturally: in other words, I'll position my mics and cables and then try my best to disappear. I'll often be given a list of dos and don'ts by the organisations I work with and I never try to bend the rules. Recording birds of prey also means being active at strange hours, and with rural crime being an issue, it pays to inform locals and even the police about your planned activities. I've often allayed the suspicions of gamekeepers by explaining that I'm another pair of eyes out there at odd times of day and night, helping them to do their job. Each Christmas, I send some of my recordings to people who've helped me throughout the year. I know some landowners who really look forward to hearing these surprising sounds recorded on their doorstep!

In terms of fieldcraft and logistics for raptor recording, things can be difficult and drawn-out. On one occasion when I was keen to record Ospreys at the nest, I got in touch with some contacts in the Scottish Highlands who recommended a suitable site. I was granted a licence, after which I assessed the site with a local ranger. To get a mic in the nest, I had to climb 15 m to the top of a pine tree with a rope and harness, then run 200 m of cable back to a natural hide. I camped in a nearby midge-infested forest, waking up every morning at 3 a.m. to start recording. Each field trip entailed a 500-mile round journey from my former home

A Northern Fulmar nesting ledge. A mini omni mic (visible centre-left) was positioned when the birds were absent, using a boom pole and a home-made wire grappling hook: an approach which enabled close-up recording with no disturbance. (Simon Elliott)

in Newcastle, but the results were worth it: I got some fantastic sounds of the female standing astride my mic, calling and feeding her young (Audio 2.5). On other occasions, I've had to abseil down to a Peregrine Falcon nest (Audio 2.6) and climb up a cliff to reach a Golden Eagle eyrie.

One of the advantages of this method of sound recording over photography is that I only need to make two visits to a nest site: an initial one to place the mic and cable, and a second to retrieve the equipment, ideally after the birds have flown. By recording responsibly and using long cables, disturbance can be kept to an absolute minimum. Sometimes disturbance is essentially zero, such as when placing a mic on a Northern Fulmar nesting ledge while the birds are out at sea (Audio 2.7). However, don't think that you're not being watched if you try this: eagles and falcons have phenomenal eyesight and will almost certainly be somewhere nearby, monitoring your every move...

Do you have any advice for constructing an effective hide? What kinds of materials and methods do you typically use?

One of the easiest hides to use is a car, since wildlife doesn't pay much attention to a human in a vehicle. It's also more comfortable than the alternatives! If I don't

Every winter, Eurasian Teal display in this Northumberland bay. The species is extremely wary: despite many hours of observation and multiple mic placements over more than 20 years, Simon has never managed to get a teal recording he is truly satisfied with. (Simon Elliott)

have a car nearby, I'll quickly construct a hide or screen from local materials such as branches, leaves, reeds and similar. Failing that, I'll use a portable, camouflage-patterned dome hide that I can assemble in a few minutes; this is one of my favourite places to be, and with an insulated mat to sit on, it's really quite comfortable.

I'm also never without a square or two of olive scrim net. These are obtainable cheaply online and offer a quick camouflage solution. I often throw one over my head to hide from face-sensitive wildfowl species, or place one over mics and windshields to break up their outlines. You can also hang one over a car window to conceal your face or body movements, and they make surprisingly good scarves on a frosty morning.

Rather than monitoring wildlife activity from a hide or other concealed location, many recordists now prefer to observe/listen beforehand, then place a drop-rig in a likely recording spot. In your own experience and bearing in mind the fieldcraft challenges involved with raptors, how does this unattended approach compare with in-person observation from a hide?

Technological developments have led to a huge rise in the popularity of unattended recording. Although I do employ this method a lot, the fundamental issue I have with it is the lack of behavioural context, or even species identification, gained through its use in the field. A recording with either of these elements missing is of little value to me. If I do use unattended techniques, I'll often use a camera trap to confirm what was going on. Personally, I get the most pleasure from being outdoors on my own and in remote habitats where few people are around, experiencing unique wildlife behaviours. I've found that if you sit quiet and still for just a few minutes, other species will accept your presence. I suppose wildlife sound recordists were 'forest bathing' decades before it became a thing!

It's also nice to have a bit of historical context when we talk about unattended recording methods. Being able to leave a high-quality recorder running for hours on end is a relatively recent development; in the reel-to-reel tape days (which I was a part of), you generally had only twenty minutes or so of recording time. Tape was expensive and you didn't waste it: you only switched things on when you were fairly sure of getting a successful take, so you had to really think about what you were doing. One of the best developments for me was the pre-record buffer. I used to spend many hours with a finger poised above the record button, only to miss the event I was waiting for. Now, with several seconds of pre-record available, it's a lot rarer to miss behaviour if you're slow off the mark. Coupled with 32-bit float recording (where managing gain and issues with clipping can pretty much be forgotten about), it's fair to say that we are in a golden age of audio technology and it's much easier to get out there and start exploring.

As an addendum to all of this, the recent and very rapid development of artificial intelligence has the potential to improve species identification, especially with unattended recordings. This is very useful and I do use AI-based programs for identification purposes, but because I don't feel it's reliable enough at the moment, I frequently cross-check AI suggestions with known recording databases. As the AI technology behind species identification improves, I may not need to worry about cross-checking as much.

Given your experience over the decades and your knowledge of how wildlife sound recording has developed, what would your main points of advice be for those starting out, especially when it comes to learning fieldcraft?

Some useful points of advice would be:

- Join a group or a society of likeminded folks. The WSRS is one of the best examples out there: it's a friendly worldwide community with over 50 years of combined knowledge and experience.
- Try to find an experienced recordist willing to take you out in the field, advise on equipment choice and help with fieldcraft and species identification.
- Get to know your subjects well, through quiet observation. For me, sitting still and watching—for as long as it takes—is fundamental to success in the field.

Getting close to wildfowl on a lake often involves deep-wading through cold water and risking expensive microphones. With care, close-miking is possible and offers a subjectively richer recording than can be obtained from the shore using a parabolic reflector. (Simon Elliott)

- When you start out, you don't have to have top-of-the-range gear. I began as a poor teenager with a cassette recorder, 50 cm of cable and a pathetic microphone! There is a certain minimum standard, of course, but I continue to make prize-winning, broadcast-quality recordings using cheap, often second-hand equipment. You can make a lot of accessories yourself (e.g. windshields) or adapt cost-effective fishing or garden equipment to your needs, without purchasing professional audio accessories.
- Design a suitable cataloguing system for your own recording library. Mine enables me to locate information on metadata such as species, call type, behaviour, habitat and time of year in just seconds, from a collection of over 13,000 sound files. This is great when providing material for film or TV, since people working in those fields usually want things delivered yesterday.
- In my opinion, you only truly learn from being out there with wildlife, experimenting with fieldcraft such as microphone positioning, how to read local topography and weather conditions, and so on. Wind direction alone has a huge effect on success or failure: one seabird location I visit is only suitable for recording on a weekday when there is an offshore wind and a low spring tide. This shows how much success depends on personal experience in the field.

In closing, I should say that in wildlife sound recording, success is never assured. I once calculated that for every ten of my carefully planned trips to the Northumberland coast to record geese, only one was successful. Even so, I never regard a trip as 'wasted'. After pursuing this craft for over 56 years, I always learn something.

Samuel Kudjodzi

Since 2021, Samuel Kudjodzi has been documenting and raising awareness of species in wetland habitats around the city of Koforidua, Ghana. This is just one example of his wide-ranging work as a wildlife sound recordist and acoustic ecologist, much of which is coordinated via the non-profit Centre for Soundscape Ecology and Multimedia (CSEM). Samuel currently splits his time between Ghana and Germany, so we focused on his activities in both countries, exploring his pedagogical practice, fieldcraft, use and modification of audio technologies, and thoughts on wildlife conservation and environmental issues. The work of Emmanuel Baffoe (a Ghanaian sound recordist mentored by Samuel) features throughout the interview, alongside Samuel's own recordings of birds and insects. *MF*

An AB stereo configuration, used to record a dusk chorus in Germany during the COVID-19 pandemic. (Samuel Kudjodzi)

How did you become interested in wildlife sound recording? What inspires you and why do you do what you do?

My earliest inspiration came from Bernie Krause's *The Great Animal Orchestra*. This work [also the title of a book by Krause] was played over a loudspeaker-orchestra system when I was a student in the Studio for Electroacoustic Music (SeaM) at Bauhaus-Universität Weimar. The collective richness of the species sounds I was hearing—and their relationship to music composition—captivated me and made me think that there was a correlation between notated music and the voices of the animal world. As a result, I decided to specialise in soundscape ecology, to help raise awareness about species and the ecosystems which make them thrive.

You have been recording quite a lot around your native Ghana, exploring the country's threatened ecosystems and natural soundscapes. Can you explain more about your work there and the wildlife you have been focusing on?

Since 2021, I have been using binaural, ORTF and AB recording techniques to document and raise awareness of amphibian and insect species in three locations around the city of Koforidua: Pantuase, Trom and Bonya. These once-thriving wetland spots are located in the foothills of Obuo Tabri, which is a famous mountain in the Eastern Region of the country. Interviews conducted there have revealed stories of formerly biodiverse rivers, with a variety of fish and bat species, birds, reptiles, crustaceans and insects once existing throughout these habitats. Community members speak of water flowing freely under the earth in the past.

Since the early 2000s, housing developments have emerged in every corner of these natural spaces. Permits have been granted for laterite and concrete to be used, to fill wetlands for human settlement. Construction activity in wetlands is growing from day to day, and concerns have been raised by the Ministry of Lands and Forestry that residential and industrial development on these sites reduces biodiversity and impacts negatively on species. The rain cycle in the area used to be constant, but now many areas are exposed to increased flooding risks because there are no wetland habitats to control surface run-off. Traditional community practices, including the gathering of fuelwood, timber, fish and water, are also lost when wetlands are converted for intensive land use.

In Pantuase, recordings made around these degraded wetland areas have revealed complex insect choruses (Audio 3.1), frog and bat activity, and some interesting perspectives on birds (Audio 3.2). Some of the amphibians recorded

Recording a wetland streamlet in Bonya, Eastern Region, Ghana. Both microphones in this AB stereo configuration have been covered with slip-on faux-fur windshields for general protection. (Samuel Kudjodzi)

there have been identified through visual analysis of the recordings, with a single spectrogram in 2022 revealing a remarkable 22 acoustic niches. This discovery raises the question: how much do we really know about the amphibians occupying these habitats? To help answer this, I'm inviting local and international amphibian specialists to assist with identification of all species heard and seen on the 2022 spectrogram. This is an arduous task, but well worth the effort for educational and outreach purposes.

A key outcome of my recording activity in these wetlands will be the creation of an immersive exhibition, along with a range of publications for schools, libraries, museums and other spaces. These resources will help educate people about the need to protect critically endangered spaces in Ghana and encourage them to be environmentally conscious when considering areas for human settlement. Better still, people may come to realise that the wetlands should remain untouched, because they provide benefits such as freshwater retention, flood prevention and carbon storage.

You are involved in educating young people about sound recording techniques and technologies, enabling them to document natural soundscapes and explore native species around Ghana. Can you tell me more about this educational work?

As part of my master's degree at Bauhaus-Universität Weimar, I developed a soundscape ecology curriculum with the help of my supervisor. Not long after this, I launched CSEM, which, amongst other things, carries forwards and implements the learning materials I developed during my studies.

When I was designing the curriculum, my main goal was to include content and resources that would educate young people about listening and teach them how to engage with the sonic environment (especially in ecological contexts, since the soundscape tells us lots about our ecosystems and the overall condition of our habitats). With the knowledge and skills gained from the learning materials implemented via CSEM, students should be able to build their own conservation-themed projects in schools and other public venues using audio technologies, photography and scientific data.

The curriculum covers competencies in sound recording/engineering, ecology and multimedia design. There are 27 subtopics, addressing themes such as:

- Ecosystems, habitats and biodiversity assessment.
- Techniques for listening to ecosystems.
- Binaural and ambisonic recording approaches.
- How to build binaural, cardioid and omnidirectional microphones.
- Monitoring, editing, mastering and sound analysis with DAWs.
- Macro photography, graphic design and visual communication.

- Multimedia terminal design and sound installation planning.
- Electroacoustic music composition and sound art.
- Sound-map development and interactive programming.

There's a lot of material, so it's taught in phases. I start with two fundamental modules: *Listening to Soundscapes without Amplification* and *Listening to Soundscapes with Amplification*. This stage of training opens up a new world of sonic realisation to students. I want them to understand that habitats can be experienced both with and without microphones and field recorders, and that each listening approach produces different points of focus. Once they are able to compare listening methods, many students are amazed at what they can hear for the first time using field recording technologies. The juxtaposed experience of listening with/without technology draws them into the craft of recording and keeps them

This binaural rig was used to record birds in Germany during winter. Earmuff-type faux-fur windshields have been attached to the artificial head. (Samuel Kudjodzi)

exploring; they become much more aware of their environment and often develop sensitive and protective attitudes towards it, which is very rewarding to see.

Other elements of the curriculum include preparing young people with skills like microphone-building, DAW techniques and the design of sound installations for public exhibitions. I also focus quite a lot on teaching binaural sound recording because of how immersive and accessible it is, and how effectively it can convey information on a sound's position and trajectory. With binaural methods, I feel that we can recreate sonic environments which are highly focused and detailed, transporting listeners directly to the habitats we have experienced in the field (Audio 3.3). I emphasise the use of closed-back headphones for installations because of their passive noise reduction qualities, and I encourage a minimalistic, budget-friendly approach to the design of sound and multimedia installations.

Scientific content is also important. I integrate data resources about environmental degradation and climate action. Where possible, I also encourage students to work with conservation scientists, bioacousticians and ecologists, so they can learn more about the habitats and species they are exploring and gain more confidence talking about scientific issues. To add a visual element to the learning, I invite experienced macro photographers to offer some advice, since understanding how to incorporate captivating images of flora and fauna helps make installations attractive and engaging.

What recording methods and technologies do you use when exploring wildlife sounds in Ghana? Have you had to make any adjustments to your recording approaches based on the climate and/or other factors?

Most of my recordings in wetland ecosystems in Ghana are binaural, because my work is typically exhibited on headphones for an immersive listening experience. However, some subjects require different configurations. I have tried ORTF and AB set-ups with cardioid mics when recording wetland streamlets, to convey a sense of texture and flow; in such cases, I found that these configurations worked very well when compared to other approaches using omnidirectional mics.

There are many different fieldcraft and technical considerations to make based on what I encounter in the field. I often find that there's a balance to be had between these two areas: I agree that fieldcraft is crucial for making good recordings, but I am also convinced that certain situations require specific technologies to carry out the task. In some cases, fieldcraft is much more important than technology; in other instances, technology is more of a focus. For me, finding a balance between these two areas depends on the challenges I encounter in the field. For example, some species can produce loud vocalisations (Audio 3.4), and it might be necessary to choose a microphone with a high maximum sound pressure level (max SPL) rating to accommodate this.

Sometimes, you just have to use what is available. Working in hot regions can be problematic for microphones, which are generally not manufactured to withstand scorching sunlight. I have found a way around this challenge by using a locally made hat (woven from rattan and often worn by market workers in Ghana) as a mic cover.

You are also based in Germany, and have recorded quite a lot in urban and suburban environments there. How does this contrast with your work in Ghana? Are there any distinct challenges that you have to work around in each country?

My intention when recording in urban and suburban spaces in Germany is to document species populations over time, and involve other people in listening so they can gain new perspectives on existing soundscapes. I sometimes set up my equipment in vantage points, with five pairs of headphones freely available so members of the public can participate in listening; this approach has helped me reach countless numbers of people, affecting how they think about the sonic environment and the species within it. I have also collaborated with media and cultural institutions in Germany to introduce field recording to young people (one such institution is the Kunstschule Wartburgkreis, which I have had the opportunity to work with since 2019). My goal through making these connections is to build up interest in the subject of sound recording and look at possibilities for integrating it into mainstream education in Germany. The response so far has been very positive and my work has caught the attention of a high-school director, so I am interested to see how things develop.

I've noticed that there are lots of mechanised sounds in Germany such as traffic, lawnmowers and aircraft (Audio 3.5). In Ghana, the sounds of religious worship (e.g. church music) and entertainment (loud music played from homes and bars) dominate the soundscape in urban and suburban spaces. Ghanaians love music and love to impress passers-by with the latest sound system, so it becomes a real challenge not to record music in calm residential areas where nature thrives. It's a very frustrating experience, when working in both Europe and Africa, to have to deal with anthropogenic noise and its impact on wildlife.

Through my participation in projects which highlight the importance of quiet spaces, I have learned how noisy environments can affect wildlife and humans by, for example, disrupting breeding patterns of species and contributing to heart disease and other health issues. In some projects I've become aware of, there is an extensive search for spaces or environments on earth where just one hour of natural sound can be recorded without hearing any anthropogenic noise. In my view, that is a very high bar to set. My experiences in the field, as well as those of Emmanuel Baffoe, one of my Ghanaian sound recording trainees, have shown that many natural spaces we assume are quiet actually have human-induced noise in them.

In Germany, I've developed a particular interest in summer soundscapes because of their rich textures and cricket melodies, which I always enjoy. I live in a suburban location where large fields contain wonderful cricket and grasshopper 'orchestras' every summer, as well as solo insect sounds (Audio 3.6). It's always amazing to me how these small insects coordinate their behaviours so well and create an other-worldly atmosphere. These kinds of things are what I am so passionate about conveying to people when I let them listen through headphones. In Ghana, the warm climate contributes to a unique soundscape of cicadas, crickets, grasshoppers and other insects (Audio 3.7), which I record during my visits. There are also some interesting bird species living in close proximity to humans (Audio 3.8). The nocturnal soundscape in suburban Ghana is surreal: the melodies of insects, tree frogs, cicadas and other animals are captivating and it is easy to spend many hours listening to them.

Attitudes to wildlife in Germany and Ghana differ. In Ghana, wildlife has not been given much focus: it is seen as being reserved for those working in conservation, and there is a general lack of funding and education on issues relating to the environment. In Germany, there is a concerted effort to document wildlife through captivating photography and other forms of media, for educational purposes. There are also very strict policies around protecting species and many institutions looking after nature are supported by the federal government. These approaches to wildlife protection in Germany have drawn me closer to the subject of conservation and I see lots of efforts to share information on wildlife in a way that invites people to participate. Conservation materials are available to all age groups and many of these are integrated into mainstream education, increasing the likelihood that children will grow up to appreciate the natural world.

Apart from training the next generation of sound practitioners, another objective of CSEM is to propose and develop partnership opportunities between a range of German/Ghanaian organisations, so we can make conservation an attractive area to work in: not just in Ghana, but in Africa as a whole. As a result of engagement activities through CSEM, I have had opportunities to share my work in scientific settings like the Fourth African Bioacoustics Conference (held in Cape Town, South Africa in 2024). My work has generated interest from the scientific community and I take great hope and comfort from this, because I believe that when science and creative practice come together, we can deliver important messages to the public much more effectively.

As an educator through CSEM and other initiatives, it must be very rewarding to pass your knowledge on to others. Do you have any thoughts on this or any student work you'd like to mention?

Having received field recording equipment from Earth.fm in 2023, I flew to Ghana and stayed there throughout October and November 2024 to begin training the first group of students through CSEM. After my arrival, I started feeling very ill and my original plan of working with five trainees at the Graduate School Auditorium of All Nations University, Koforidua didn't work out. Nonetheless, I was able to give recording advice to Emmanuel Baffoe (mentioned previously) throughout my recovery. I covered a range of topics with Emmanuel, such as monitoring techniques, focused/peripheral listening, how omnidirectional microphones work, microphone placement, and so on.

In many respects, the CSEM programme was piloted by Emmanuel, whose recordings have featured throughout this interview. He was the first student to use all of the educational materials I developed and he will now begin passing on his skills to others. He has progressed very quickly, demonstrating an independent learning approach since continuing his mentorship with me remotely. Emmanuel has also started sharing his knowledge with a female practitioner who will be spearheading a new project called Women In Soundscape Ecology Africa (WISEA): an educational programme aimed towards female students and working professionals.

Recording amphibians at midday in Ghana, employing an artificial binaural head. Note the soft cover on the field recorder (extreme foreground), to protect the unit from direct sunlight. (Samuel Kudjodzi)

In this programme, girls and female leaders will be mentored to acquire skills in sound engineering, soundscape ecology and multimedia presentation, which will help them to build their own conservation-themed projects in the future.

At the moment, Emmanuel is recording in fragmented wetland sites in Koforidua, adding to existing binaural archives gathered there by CSEM between 2021 and 2022. His work will be used for an upcoming installation on wetlands, detailing their positive effects on biodiversity and human life. He has also embarked on a field recording project exploring critically endangered ecosystems, encompassing Ramsar sites in Ghana as well the Atewa Range Forest Reserve and Bobiri Butterfly Sanctuary. These biodiverse ecosystems are crucial for the survival of endangered species, but sadly, they are now at risk of mining and human settlement.

Emmanuel is facing funding issues for the project and is working with very limited resources, so I can only hope that he receives as much support as possible to continue his valuable work and help raise awareness of Ghana's wildlife.

Emmanuel Baffoe, recording in a wetland. Emmanuel has recently embarked on a field recording project exploring critically endangered ecosystems in Ghana. (Justice Ayertey)

Jo Kennedy

After working in the UK environmental sector for over 20 years, Jo Kennedy took a bold new career direction: one focusing on wildlife sound recording's creative possibilities and its potential for articulating our relationship with the natural world. I first encountered Jo's work via the *Nature Tripping* podcast (co-presented with her partner Cathy Shaw), which features many of her recordings and explores a range of interrelated community and conservation topics. In this interview, she discusses fieldcraft and technical approaches to the podcast, as well as childhood influences, her career transition, and the relationship between wildlife sound recording and creative practice. She also talks about gender inequality within the audio industry, offering a positive glimpse into her awareness-raising activities in this area and words of encouragement to budding sound professionals. *MF*

Jo sits by a mountain lake near Aulus-les-Bains in the French Pyrenees, listening with a hydrophone. (Andrew Bailey)

How did you become interested in wildlife sound recording? What inspires you and why do you do what you do?

I think I record wildlife because it involves listening, and I feel that listening is one way of arriving at a deeper understanding of the natural world: of having a relationship with it and making friends with it.

I like existing on this planet alongside other species. I like going out and seeing or hearing who's there, exploring and wondering what other creatures are up to. You learn a lot about nature through trying to record it, and I really appreciate that: I love the closer connection to the 'more-than-human' world that sound recording facilitates. As to why I feel compelled to record wildlife (and considering why listening in itself isn't enough for me), I suspect some of that is down to the enjoyment of engaging with the technical challenges of making a successful recording, and all the decisions and lucky moments that go with it. There's human curiosity there too, interwoven with the slightly egotistical satisfaction we all get from showing other people what we have experienced.

You studied environmental science, which led to a career in the UK environmental sector for over 20 years. You then decided to study Creative Music

Technology at the University of Huddersfield. What prompted this transition and how did it encourage your interest in wildlife sound recording?

My decision to move away from the environmental sector towards working with sound came from two sources: one was dissatisfaction with my job, and the other was an internal desire to have a more creative life. Together, they manifested as a kind of existential anxiety which bubbled away and eventually precipitated a jump.

As strange as it may sound, I was finding it more and more difficult as an environmentalist to work for the environment agency I had built a career in. You might think such an agency would be an ideal place for someone passionate about nature and wildlife, but by 2014 (when I left), my employer had undergone successive 'change programmes' which meant that fewer resources were available to do my job properly. Our policy-making powers as an agency had been weakened and there was a deregulatory push underway by the UK government at the time. It felt like the bold ambition of previous years had vanished. I looked at the time and energy my job was costing me compared to what I was actually achieving and felt increasingly disillusioned about the value of my work.

Meanwhile, I had a creative side. I had been writing songs, taking online courses in composition and was thinking of doing a Music degree. When another work reorganisation came around, I finally decided to make the leap. I began studying Creative Music Technology at Huddersfield, where conventional musical genres such as classical, pop and others were on the curriculum; but there were also opportunities to explore more experimental avenues—to compose with sounds rather than music notation—and it was this area that truly fascinated me. All of a sudden, field recordings, encompassing sounds from landscapes, the weather and wildlife, became valid compositional sources. I realised that by using a small handheld recorder, it was possible to go out and collect sonic materials in the field and start working with them creatively.

Does your background in environmental science inform your creative practice in any way? Or is it something that you have kept separate, as a former career path?

When I was little, I spent hours outside on my own, exploring the railway embankment on the other side of the garden fence. I poked anthills, climbed trees, and observed insects and birds. I loved looking for spiders, moths, rats: things that repelled most other people. I think I've always been attracted to the natural world on a deeply personal level.

This fascination with nature informed my decision to study environmental science, and my later ventures into recording habitats and species. All of my work shares the same roots, drawing from my interest in experiencing the world around me and understanding it through an ecological lens. Today, my environmental science background and my interests in audio inform each other; that said, I'm more connected to the living world as an audio practitioner and sound recordist

Jo descends from Bleaklow Moor, Derbyshire with Carlos Bedson, having just finished recording for a *Nature Tripping* podcast episode on Mountain Hares. (C. M. Shaw)

than I was in my previous career, since a lot of work in the environmental sector involved sitting behind a desk and going to meetings.

The projects I'm most interested in, whether as part of my own creative practice or when working for others, are those that explore and question our relationship with the living world. I like the notion that art might be able to nudge people into pro-environmental behaviour and I'm happy to produce work about ecological issues. I think art has greater potential than facts and figures to evoke an emotional response in people and engender a sense of care for nature, which can lead to meaningful and positive changes in behaviour.

Your podcast, *Nature Tripping*, involves heading out on short trips to record wildlife and explore the natural world. How do you approach a typical episode?

Nature Tripping is about connecting people with nature through experiences in the field. I co-present it with my partner Cathy. We think of an idea, do some research and draft a structure for each episode before we meet with any featured guests or experts.

In terms of equipment, as a bare minimum and if I only have enough time on-location for an interview with a featured guest (i.e. if there are no additional recordings of wildlife or habitats), I'll use a stereo pair of mini omnidirectional mics mounted inside a basket windshield. I place this rig on top of a microphone stand and always keep a furry cover on the basket, since I've learnt from experience that there's never enough time to apply one when the wind picks up. It's worth mentioning that I bought all of this windshielding gear second-hand in an audio shop for just £50, after explaining what I was trying to do. They just had it lying around in a back room! To enable stereo recording with this second-hand set-up, I secured a length of wooden dowel down the centre of the basket, using the remnants of the internal shock-mount (originally intended to hold a shotgun mic) as a support; I then taped a mini omni mic to each end of the dowel, creating an AB stereo pair.

Taking location advice from your interviewee(s) is recommended, as is doing a bit of reconnaissance to find the best recording sites. Wildlife doesn't just turn up when you want it to, and species may have song-posts or established territories where you are more likely to encounter them at certain times of day/night. I'll take note of any landscape features I can use to help with windshielding or traffic noise absorption, and I'll try to build a mental picture of nearby watercourses, which can be quite noisy and mask people's voices during interviews. I usually also take some time to listen to a recording location over an extended period, to understand how everything sounds through headphones. Listening to an environment is actually one of the most important elements to remember: sounds have incredible potential to transport people to a particular place, and you

Recording overwintering Eurasian Curlews and other wetland birds as the tide comes in at Morecambe Bay, Lancashire. A stereo parabolic reflector set-up is being used to focus on distant sounds across the water. (C. M. Shaw)

need to understand how the listener will experience things so you can do justice to the species and habitats there.

I don't always manage to, but I like to collect additional recordings which can be woven into the podcast or provide sonic interludes between spoken word sections. If featured guests are discussing a habitat with me, I might decide to use an AB pair or an ORTF configuration for this (Audio 4.1); if they are discussing a species, I'll bring my parabolic reflector to try and record some point-source detail (Audio 4.2). I've also used a stereo configuration in my parabolic reflector to record ambiences (Audio 4.3). Lapel mics at close range (e.g. taped to grass stalks) are a good option for bumblebee nests: very discreet and easy to deploy. Ultimately, I've found that one size or approach never fits all, and you have to choose how to tackle things based on circumstances and the podcast topic in question. If you are working on foot, there is only so much equipment you can carry, so some upfront decision-making about what you want to focus on really helps.

Are there any wildlife encounters that have stood out for you in the podcast so far? Any special moments or notable challenges?

Making the podcast is a lovely way to explore nature. Every episode deepens my understanding of the living world and our relationship with it.

Some parts require a lot more preparation and recording time than others. One such example is an episode we did on Common House Martins (Audio 4.4). The joy of this one was the human angle we explored: two people—Jill and Kathryn—who went from wondering what was leaving a messy 'pile of poo' on their terrace, to realising that house martins had arrived for the summer and were making their homes under their eaves. They experienced close-up perspectives, observing, wondering and learning what the birds were up to and building an emotional connection with them, riding the ups and downs of the breeding season.

The house martins' nest was under the eaves of a tall Victorian stone house situated close to a bustling town centre and a river, so the area was challenging to work in from the beginning. After observing the birds from the garden terrace, I realised that airborne passes and individual visits to the nest were momentary and sporadic: if I wanted to record the species' intimate chattering noises effectively, the equipment we had brought would need to be left running for extended periods. I gave Jill a short lesson on how to operate a handheld recorder, to which I connected an affordable pair of mini omnidirectional mics. After adding furry windshields, we positioned the mics outside, securing them either side of a sash window-frame close to the nest. I asked Jill to turn the recorder on and leave it running when she noticed any activity.

During production of this episode, we also discussed and included the sounds of other species (related hirundines such as Barn Swallows and Sand Martins, and the unrelated Common Swifts), and talked about how to differentiate between them visually and sonically. I ended up incorporating some Sand Martin sounds from a colony I had explored on the Hebridean island of Tiree in Scotland. My recording method for the Sand Martins was not exactly ideal, but it worked. On a morning with low winds, I snuck into sand dunes as carefully as possible to avoid disturbing the colony, and lay on my back in the soft, damp sand which had spiky marram grass poking through it. I held my parabolic reflector as close to my chest as I could for stability and pointed it upwards, using my headphones to guide the angle I was holding it at. I stayed there for a while. It was quite uncomfortable, but it was wonderful to close my eyes and immerse myself into the coastal soundscape as I recorded the birds passing overhead.

One of the best scenarios to find yourself in during podcast production is when you engage an expert who is knowledgeable about the landscape you are covering and who is willing to take you to the best spots at suitable times. We were lucky on one occasion to have the expertise of National Trust ranger Darren Mason, who took us across Sandscale Haws in the Lake District on an April evening, in search of the UK-protected Natterjack Toad. As darkness fell, the quiet coastal ambience gave way to a chorus of toads, each calling for a mate (Audio 4.5). It was an incredible transition. There's a main road in the background of the recording, but the juxtaposition of the toads and human-generated sounds serves as a meaningful reminder that we need to leave some space for coexistence with wild species.

A stereo rig, positioned near a pond at Roanhead, Barrow-in-Furness in preparation for Natterjack Toad recording at sunset. The Natterjack Toad is a protected species in the UK and a National Trust officer was present at all times. (Jo Kennedy)

Creating our podcast episodes has generally been fantastic fun; that said, I've had a few moments of jeopardy, including one scary experience with cattle when I was trying to record bumblebees. Despite their docile reputation, you should always be respectful of cattle, especially when they have young nearby since the mothers can become quite defensive. I also had a recent confrontation with a landowner when recording Eurasian Curlews, which left me quite shaken. As a recordist, it's easy to follow your microphone wherever it takes you, but you need to be prepared for the unexpected and keep personal safety in mind too.

I'm interested in the relationship between wildlife sound recording and your creative practice. Do you view them as one and the same, or as separate activities?

They are separate but interrelated.

Sometimes I go recording purely for the sake of it. I love the skill involved, the occasional sense of adventure as I walk alone through remote spaces, and the shift in headspace it affords. It encourages patience and requires you to be still, remaining in the moment. I nearly always feel calmer and more grounded after being outside, listening and recording.

There are occasions when the podcast or a creative project requires wildlife sound recording and I have target species/habitats in mind. For example, I recently worked on a collaborative audio-visual installation celebrating the local Eurasian Curlew population. For that, I needed to be present around the estuarine and moorland landscapes which curlews inhabit, so I could gather sounds with emotional context and make something that was authentic. Weaving these requirements together can be very taxing, but I like the fact that I can create work out of the recordings I've made: work which might inform, engage and connect people with the living world. It's a huge source of motivation for me as a sound artist.

You are a member of the UK-based Yorkshire Sound Women Network (YSWN). What kind of work does YSWN do, and what inspired you to join?

I'm involved in YSWN because I want to challenge gender inequality in the audio sector. At university, I saw stark imbalances: textbooks and lecture materials only citing male examples and work; a cohort of overwhelmingly male music technology students; and female students being pushed to the sidelines in subtle ways. If you go online and look at audio company profiles and marketing materials as well as forums and videos on music technology, you often see an overwhelmingly male world. Because this kind of gender stereotyping is so deeply engrained in our culture, everyone misses out on a more interesting and diverse society. If interesting work is done by people outside the cultural mainstream (especially audio work), it's often the case that nobody is aware, because by and large, mainstream culture doesn't encourage an atmosphere of support or promotion.

Whichever niche of the audio industry you look at, evidence from surveys and studies shows that the situation is stacked against people who aren't white, male and able-bodied. Girls frequently make it through a male-biased education system only to enter workspaces and working cultures designed by and for men: as a result, they struggle to find anyone to relate to. At YSWN, we hear lots of real-world examples: female audio professionals turning up on film sets and assumptions being made that they are either make-up artists or cleaners; female sound conference attendees finding that men only address other men, or only comment on what women attendees are wearing rather than ask about their work; and female performers having to deal with male sound engineers who assume they know nothing about audio equipment and ignore their wishes on how things should sound.

The extra energy it takes to navigate this world daily means that talented people leave creative spaces which would otherwise be fantastic to work in. YSWN is one of several initiatives that have sprung up to try and change this. We are about providing role-models in the audio sector for girls, women and people of minority genders. We help individuals develop their skills and build confidence through creative commissions, work placements, workshops and seminars. We support each other as a community through networks and events. We also carry out research and use this to provide training—for both educational establishments and the audio industry—on how to create more equitable teaching environments and workspaces.

Women are more at risk when working solo in remote locations, but cities can feel every bit as intimidating in terms of personal safety. Having said this, I was speaking with Melissa Pons recently (another wildlife sound recordist), and she pointed out that one of the advantages of being outdoors in the field is that you can be pretty self-sufficient: you can do your work away from spaces where there are issues with unconscious gender biases, toxic masculinity and so on. I've found some great role-models and inspiration from women field recordists such as Melissa. Whilst we can work to lift each other up, it's also up to those who commission work—organisations, institutions, funding bodies, businesses and male allies—to be more inquisitive, imaginative and mindful of gender equality. They can seek us out and engage with us, and we can all work together to build a diverse and equal recording community.

What advice do you have for recordists starting out on their journey, especially those considering a career transition?

I deliberated for a long time before changing my career path. In hindsight, it may have been better to make the jump earlier, but that's a very personal decision based on individual circumstances (including the ability to take financial risks!). Don't be too hard on yourself if you find that you're sitting on the fence.

Also, bear in mind that any process of transition is rarely a straight line; when I moved into audio, for example, I didn't realise there were so many niches within

A home-made binaural head, left for an extended period to collect ambience at Little Woolden Moss, a lowland blanket bog habitat on the outskirts of Manchester, UK. This site was once exploited for peat extraction but now undergoing restoration by the Wildlife Trusts. The artificial silicone ears of the binaural head have been covered by faux-fur windshields. See Audio 4.6 for an example of a recording made using the pictured set-up. (Jo Kennedy)

the sector. I've tried several different types of audio work under the umbrella of 'working creatively with sound' and am still iterating as I go along. Wildlife sound recording and ecological sound art are my passions, but I'm active in other areas such as podcast production, live sound (e.g. musical performance) and workshops.

If you're going to make the jump, it's really important to have supportive people around you: people who are on your side whilst you find your feet in a new sector, since there will inevitably be ups and downs. If you are naturally introverted like me, you need to gather up your courage and put yourself out there, whether that's through a website, social media or both. And if you are aiming to go freelance, one difficult transition is to go from being fairly sure you can do the work because you have built up enough skills and knowledge, to having enough confidence to pitch for bigger jobs and charge professional fees. Job-shadowing opportunities and chats with people in the industry can be invaluable in that regard. Don't forget to look out for talent development schemes too: not all are for younger people. It's also very useful to join societies or organisations like the WSRS and YSWN.

Finally, carving out a successful path for yourself isn't all about making money. My own unpaid personal projects, like the *Nature Tripping* podcast and a DIY binaural head I built and wrote a blog about, helped showcase what I do with sound and gave people confidence to hire me. So, my final piece of advice is to get out there and make something!

Greg Green

In his day job as an audio cataloguer at the British Library (BL) in London, Greg Green works with extensive collections of historical wildlife sound recordings and their accompanying metadata (including spoken field notes and written/typed documentation), with the aim of making these valuable materials more accessible to researchers and the wider public. In addition to focusing on this fascinating line of work, we discussed his recent trips to Sweden and Mongolia, fieldcraft approaches, broader goals as a recordist, ethical considerations, and passion for sharing knowledge with new generations. In his last response, Greg offered a few points of advice for younger wildlife sound recordists. *MF*

How did you become interested in wildlife sound recording? What inspires you and why do you do what you do?

Recording satisfies my obsession with collecting, my desire to preserve memories of places and things, and my passions for nature and sound.

I developed a love of wildlife from early childhood, particularly birds. My parents have a beautiful garden in Kent, and I spent many childhood hours there looking for different species (I still do, whenever I visit). They taught me to identify birds and would take me and my brother to nature reserves all the time. In my teens, the pressure of trying to fit in at school—where being a nature nerd wasn't 'cool'—led me to bury my passion for the natural world just a little. At that point, I got into music, picked up the piano and dreamt of playing keys in a band.

In my late teens, I got an A-level [UK school/college qualification] in Music Technology, then went to study Music Production at Leeds Beckett University. The course covered the audio industry in significant detail and gave me access to all kinds of recording gear. In my final year, I enrolled on an elective field recording module; I was instantly hooked, and began borrowing mics and other equipment from the university to record around parks and woodland. During the module, Chris Watson delivered a guest lecture in which he talked about his field experiences/approaches and played some multichannel recordings of wildlife. That was the moment when I decided to fully immerse myself in wildlife sound recording. I joined the WSRS and developed my skills at field meetings (Audio 5.1), interacting with and learning from likeminded people. My passion grew and grew and I've not looked back.

You currently work as an audio cataloguer at the BL. Can you tell me about that role and how it informs your wildlife sound recording practice?

Just as I was completing my undergraduate degree, several jobs came up in the BL Sound Archive. They were all part of something called *Unlocking Our Sound Heritage*: a five-year project aimed at digitising, cataloguing and making accessible hundreds of thousands of unpublished sound recordings. These recordings covered all curatorial areas, not just wildlife, and up to that point they had only existed on physical carriers like open-reel tape, cassette, MiniDisc, DAT and CD-R. Even in controlled archival conditions, these physical media will gradually deteriorate.

I'd already taken an interest in the BL Sound Archive while working on a field recording project at university, and I'd had some interaction with Cheryl Tipp, the curator of wildlife sound. She told me all about the library's digitisation project and encouraged me to apply for a cataloguing job, which I was successful with. When I started working with the collection in 2017, I was cataloguing a whole range of materials, but with the support and mentoring of Cheryl and my other wildlife sound cataloguing colleague (the aptly named Paul Duck!), I soon became one of the wildlife sound specialists. At the time of this interview [January 2025], I've been working at the BL for seven years.

My job as an audio cataloguer involves preserving all of the useful details relating to physical carriers and digitised recordings, with the aim of making these valuable sounds and their accompanying metadata more accessible to researchers and other interested parties. For most of these details, I rely on information documented by the recordist. This can mean reading through notebooks and spreadsheets, listening to field slates, or even looking at scribbled notes on a cassette-tape box. Ideally, I'll want to find info on things like: species; date; location; time; recording equipment and notes on how it was used; weather conditions; habitat notes; and behavioural context. I can't always have all of those elements, but if I get at least species, location and date, that constitutes a good catalogue record.

Most of the wildlife collections I've come across are extremely well organised, which seems to be a typical of people involved in the wildlife sound recording craft. Take the collection of late WSRS members Charles and Heather Myers, for example (Audio 5.2, 5.3). They made thousands of recordings on quarter-inch tape and took the time to splice every single one onto reels arranged by species. Between every cut, they included a length of leader tape on which they wrote subject, date, location and equipment info. They also provided ring binders filled with recording information on every cut, ordered in the same manner as the tapes. Richard Margoschis (one of the founding members of the WSRS) also had a meticulous approach to cataloguing, and I suspect his organisational rigour may have inspired Charles and Heather's. One of my favourite recordings of Richard's from the BL archives features a Carrion Crow in winter woodland, accompanied by an eerie soundscape as wind whistles past the microphones and through the bare trees (Audio 5.4).

In addition, the BL has a huge collection of species reels featuring recordings of wildlife from all over the world, contributed by hundreds of recordists. These tape reels are organised in a very similar way to Charles and Heather's but on a much larger scale. Every contributor would have had to fill in a cataloguing template sheet, and those sheets, like the tapes, were all kept in order of scientific species

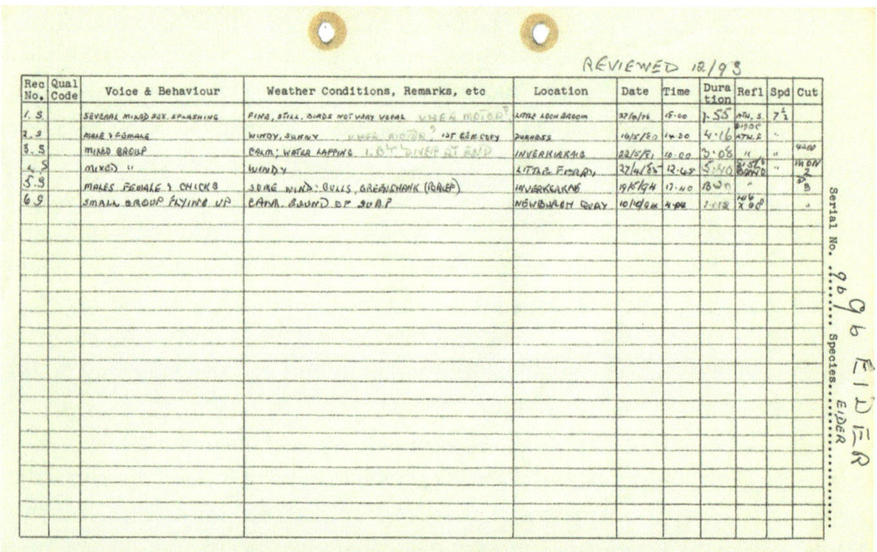

REVIEWED 12/93

Rec No.	Qual Code	Voice & Behaviour	Weather Conditions, Remarks, etc	Location	Date	Time	Duration	Refl	Spd	Cut
1. S.		SEVERAL MIXED SEX. SPLASHING	FINE, STILL. BIRDS NOT VERY VOCAL [illegible]	LITTLE LOCH BROOM	27/4/76	15.00	1.55	ATH. S.	7½	
2. S		MALE & FEMALE	WINDY, SUNNY [illegible] 1ST GEN COPY	DURNESS	16/5/80	14.20	4.16	[illegible] ATH. F	"	
3. S		MIXED GROUP	CALM; WATER LAPPING [illegible]	INVERKIRKAIG	22/5/81	10.00	3.08	"	"	[illegible]
4. S		MIXED "	WINDY	LITTLE FERRY	27/4/85	12.45	5.40	[illegible]	"	MON 2
5. S		MALES FEMALE & CHICKS	SOME WIND; GULLS, GREENSHANK (BRIEF)	INVERKIRKAIG	19/5/94	17.40	[illegible]	"		B
6. S		SMALL GROUP FLYING UP	CALM. SOUND OF SURF	NEWBURGH QUAY	10/10/94	4.pm	1.12	[illegible]		"

Serial No. 96 Species EIDER

96 EIDER

An example of the cataloguing approach of Charles and Heather Myers, from their personal collection. See Audio 5.2 for the recording noted in the fifth row of the pictured documentation sheet. (Greg Green / British Library)

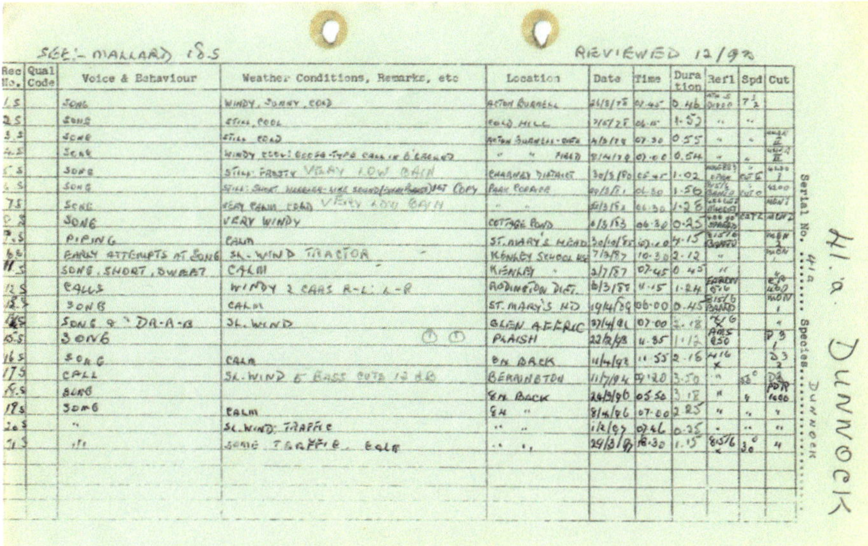

A more populated species example of the cataloguing approach of Charles and Heather Myers, from their personal collection. (Greg Green / British Library)

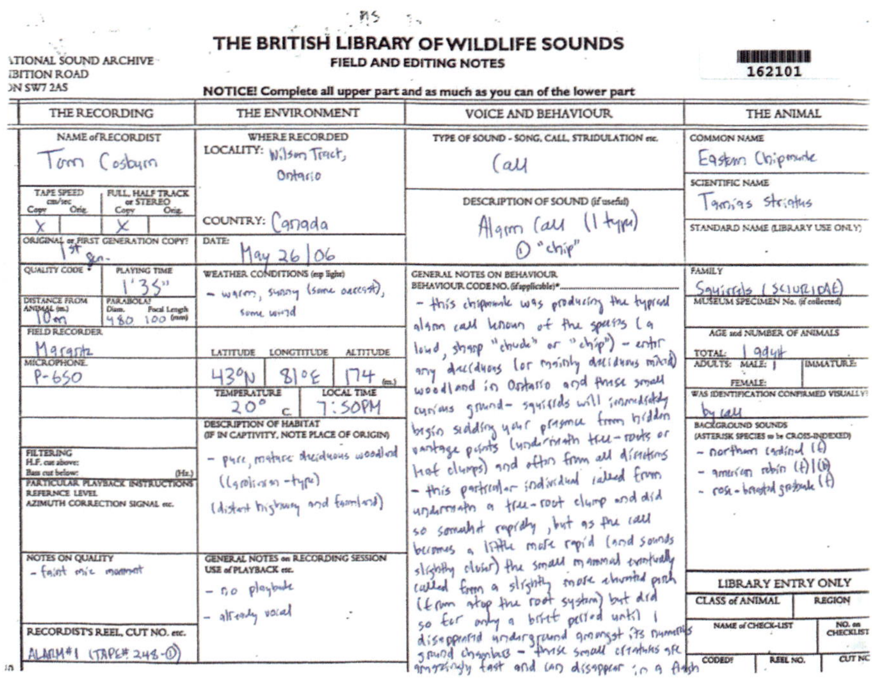

An example of a well-documented BL wildlife sound cataloguing sheet, filled in by the contributing recordist, Tom Cosburn. (Greg Green / British Library)

Some of the over 7,000 wildlife species reels shelved in the basements of the BL in London. (Greg Green / British Library)

name. Even if a recordist hadn't catalogued in great detail, they would often add fairly detailed slates in the field, which gives me enough context to create valuable catalogue entries.

In terms of how the job has informed my own practice, you might think I have a meticulously organised database of all my recordings. I wish that were true and I want to make it so eventually, but it's not the case. Because of my job, I know what information is important and I try to document as much of it as possible; I attach typed notes along with the audio in my computer's file browser, and within these notes, I include species, location, date, time, equipment and any other info I think might be useful. Maybe one day I'll have everything in a proper searchable database…

It can take me a long time to catalogue and edit my recordings. I have a perpetually filled folder called 'unsorted', where everything gets dumped until I get to it. After spending all week cataloguing other people's collections, it can be hard to motivate myself to look at my own, so I always try to slate important information at the end of my recordings: enough to jog my memory later. Using a smartphone, I also take photos of my recording equipment and the habitat I'm working in, so I can refer to these images for GPS coordinates and other useful details.

You do most of your recording in the UK, but you've also been on a recording trip to Sweden. How did Sweden compare to the UK with regard to fieldcraft? Were there any challenges you hadn't come across before?

Prior to going to Sweden in 2022, I hadn't recorded outside of the UK. It was also my first time flying on a plane since I was a toddler, so it was an anxious experience. Despite this, it felt like a soft introduction to recording in a different location, since much of the wildlife there was familiar. I went to a region called Svartådalen, about 130 km north-west of Stockholm. It's a vast expanse of coniferous woodland, rivers and lakes, encompassing the Hälleskogsbrännan Nature Reserve (which was the site of a devastating forest fire in 2014).

One of the first things I noticed in Sweden was how quiet things were. Spending all of my time recording in the UK, I had got used to almost constant background noise such as road/air traffic and farming activity. There was still human-produced noise in Sweden, of course, but it wasn't constant; when vehicles or machines were heard, they pierced through the relative silence rather than adding to a collective background din. In mid-April, the climate wasn't that different to the UK's: just a bit later in the seasonal shift. There was a significant amount of snow on the ground when I arrived, but by the end of my week there most of it had thawed and daytime temperatures reached 20 °C or more.

I went with a group led by Stefan Taylor, who owned a cabin and did ecological work in the region. He had a licence to enter Western Capercaillie and Black Grouse breeding grounds, and he took the group over to these sites outside of the birds' active times to set up some low-profile drop-rigs. I was very excited to have this opportunity, since the Western Capercaillie had been a target species for me since I started recording. Stefan's knowledge of capercaillie leks, combined with his use of trail cameras, helped us decide where to set everything up. On my first visit, I got decent recordings of lekking but they weren't especially close; on my second visit, I repositioned everything and got much better results.

I returned to Sweden in spring 2023. On four subsequent visits to the same capercaillie site, I deployed my own camera trap as well as my mics. On each recording attempt, I fine-tuned my set-up based on the observed habits of the male bird. I wanted a stereo recording that documented the space and movement of the lek, as well as the intimate details of the bird's vocalisations. Eventually, I got

Greg sets up mics at a Western Capercaillie lek in Sweden, using the low branches of a pine tree for cover. (Alan Burbidge)

what I was after (Audio 5.5). There are moments when the male moves away and you can hear the sound reflect off the trees; then he comes in close and you can hear the vibration of his feathers and even his breathing as he calls out.

Mongolia was also a recent recording destination for you. It was a fresh experience and presumably very different from your trips in the UK and Sweden, so I'm curious about how you approached things.

Mongolia was very different to anything I've ever done before, and a big step up from Sweden and the UK. I went in 2023 with a group of recordists I'd met through the WSRS. Patrick Franke, our organiser, liaised with Bolormunkh Erdenekhuu (a Mongolian wildlife tour guide), to create an itinerary suitable for a group of sound recordists. Bolormunkh was accustomed to leading groups of photographers and birders, so this was a different experience for him; nonetheless, he was very accommodating and flexible, showing us around, pointing out species and letting us go off to record by ourselves. We travelled in old Russian off-road vans, all the way from Ulaanbaatar [the Mongolian capital] to the north-east of the country over the course of two weeks. At each stop, the guides and drivers set up our campsite and we were provided with three very nice meals a day. Considering how remote everything was, it felt quite luxurious. The only things I really missed were a proper shower and toilet!

We were quite lucky with the weather when we went in June. It was mostly clear and sunny during the day. Temperatures fluctuated from single figures at night to 25 °C or more in the afternoon; just after sunrise, I went out with multiple layers on, but by mid-morning I'd usually be down to shorts and a T-shirt. It didn't rain too much, although a couple of days were very wet. In Mongolia, the habitat is mostly open grassland with very few trees, interspersed with wetland zones, riparian forests and rocky hills. Due to the exposed nature of the landscape, wind was a challenge on some occasions: I had to choose my mics and windshields carefully, seeking out sheltered positions behind trees and rocks and exploiting dips in the landscape.

I didn't go with target species in mind, so I packed my gear with a generalist approach. I was able to take two stereo rigs (mid-side and SASS-type), a lightweight parabolic reflector, a small selection of omnidirectional mics and two field recorders. That was enough to allow quite a bit of flexibility; I could deploy one set-up as a drop-rig and head off elsewhere with a reflector or a second stereo rig, for example. I also took a laptop and external SSD for backing up and cataloguing in the field.

Lack of electricity was something I had to plan for, so I checked in advance with airlines to establish which batteries I could take. I was allowed to carry: two power-banks, each with a maximum 100 watt-hour capacity; two rechargeable lithium-ion batteries; and some AA batteries installed in all my devices, as a last resort if everything else died. I also brought a relatively inexpensive foldable solar charging pad, as did the other people I went with; whilst not very powerful, having it meant that depleted batteries could be left charging all day, which was a lifesaver.

A dilapidated livestock pen in the Mongolian steppe, taken over by nesting ravens, choughs and Tree Sparrows. (Greg Green)

Greg points a parabolic reflector towards a Black Woodpecker nest in Mongolia. (Clemens Pitschke)

The old Russian vans also had charging capabilities via dashboard cigarette lighters, which were fought over regularly by five desperate recordists!

I was opportunistic in my approach, spending most of my time with a parabolic reflector in hand, ready for anything. There were times when I heard or saw something I wanted to pursue. For example, one campsite was beside a stretch of riparian habitat with a dead tree sticking out on one side; I saw that tree as a potential song-post, so I stuck a mid-side rig in the undergrowth with the mics pointing up at it. I made a good recording of a Siberian Rubythroat that way (Audio 5.6), which I'd tried and failed to record with a parabolic reflector. There were also raven, rook, chough, sparrow and woodpecker nest sites, which provided ideal drop-rig opportunities (Audio 5.7).

How do you decide on a particular recording method (e.g. parabolic reflector, SASS-type, drop-rig)? Are your choices based on the species you'd like to focus on, or other factors such as equipment weight/bulkiness or potential disturbance to your subjects?

Recently, I've been leaning towards lighter, easier-to-use configurations. I made my SASS-type rig at home using yoga blocks, cheap omnidirectional mics and a sheet of furry material for wind protection: all very portable and convenient. On a recent trip to Islay in Scotland, almost all my recordings were made using a lightweight parabolic reflector with a small lavalier mic mounted at the focus,

A home-made SASS-type rig, made from a yoga block, a pair of tights and some faux-fur windshield material. (Greg Green)

and a low-profile portable recorder clipped to the handle (Audio 5.8). Other than headphones, that was it! No tripod, no big shoulder bag or reel of heavy cables. Using a 32-bit float recorder meant I didn't need to worry about setting gain levels. I could have got better results with different equipment, but this lightweight approach was incredibly freeing and allowed me to feel much more immersed in the environment.

I've done plenty of work with drop-rigs. It sometimes feels like a lazy approach, but they're great for scenarios where being in the field physically would be too difficult or create disturbance. As discussed earlier, I often leave a drop-rig out while I go elsewhere with a second set-up; a small handheld recorder with a stereo pair of mics clipped to a tree trunk or branches is perfect to leave unattended in these scenarios. That said, there are still challenges and disadvantages associated with drop-rigs: you put your gear at risk of theft or damage from humans or wildlife, and even without human presence, equipment can cause disturbance or change animal behaviour. For example, during one long-duration unattended recording attempt in my garden, a bird started plucking faux fur from my windshield to use as nest material. I wish I had done more to prevent that!

Long recordings made with unattended rigs also take lots of time to sift through and use a lot of computer storage when they're unedited. Like most other folks, I use a combination of spectrogram and waveform checks to help find moments of interest in audio files, but this still takes time and I'm sure I miss some interesting things when scanning through material. With drop-rigs, there is no observational context for the sonic event and it's important to remember that this can be vital to understanding and documenting species behaviour.

What are your long-term goals with wildlife sound recording? Are there any fieldcraft techniques you'd like to develop or refine, or elements of the craft you'd like to explore in greater detail?

There are always more species that I want to record. I'd like to get more terrestrial mammal recordings, since they present their own unique sets of challenges in terms of fieldcraft. I'd also like to try multichannel recording: it's an area that has always excited and interested me, but I've never had the time, money or space to get into it.

Going forward, I really want to use my recordings and field experience to benefit people and the planet. It's such a privilege to be able to go out into nature to record these sounds, and I feel a debt is owed for this. I recently had the opportunity to teach a few lessons in a school close to my hometown and it was a very rewarding experience; the kids were really engaged and it's definitely something I want to do more of. As it happens, the school backs onto the National Trust's Knole Park, where I made one of my earliest recordings of Fallow Deer during my university studies (Audio 5.9). It was a real 'full-circle' moment, returning there and talking to the next generation about the wildlife sounds on our doorstep.

I want to use recordings to positively benefit the wildlife and places I have experienced. I see my focus remaining mostly local in the future, working with people, organisations and reserves in the UK. I don't like to travel too much for a number of reasons—carbon emissions being one—and I think there is a lot of positive impact to be had through working in local environments, with local people. That said, the state of nature in the UK is dire: it can feel like a sinking ship at times, and I can only hope that my feelings in this regard will change for the better.

Following on from the previous question, do you have any advice for younger people starting out on their wildlife sound recording journey?

It might be an obvious thing to say, but the welfare of wildlife should always come first. Mistakes will likely happen when you're starting out, but everything should be done with other species and their habitats in mind. Learning to identify signs of stress like alarm calls will give you an indication of when you're causing disturbance. It goes without saying that you shouldn't go for any rare or protected species early on, since most will require a licence to pursue legally. Focusing on common species is the best way to learn.

Also, of course, as someone who catalogues recordings professionally, I'm obliged to say you absolutely must write down or announce every single detail about a recording on the assumption that somebody will, someday, be searching for it!

Other than these things, the best advice I can give is to share your passion with other people, both inside and outside of the wildlife sound recording community.

Greg, standing in front of the *Animals: Art, Science and Sound* exhibition at the BL. (Catherine Smith)

Things like equipment choice, fieldcraft techniques and species identification can seem daunting, but there's a lot of camaraderie amongst wildlife sound recordists and there's always someone happy to pass on knowledge and expertise.

Nikki Sheth

Beginning with childhood influences in the garden and ending with words on diversity and equality, in this interview, Nikki Sheth discusses her academic research and creative practice, touching upon recent trips to South Africa and Australia and explaining how her wildlife recordings have been used in a range of electroacoustic compositions and sonic arts projects. Her recordings and excerpted creative works provide an audible backdrop to our discussion, as she considers species such as the Common Hippopotamus, Eastern Whipbird and Grey-headed Flying Fox, the biodiverse habitats she works in, and how sounds from the natural world weave through her practice in the field and studio. *MF*

How did you become interested in wildlife sound recording? What inspires you and why do you do what you do?

I've always loved sounds from the natural world. I remember spending hours in the garden with my grandma when I was little, listening to wildlife, gardening and using bird whistles, so I feel that I've always had a connection to sounds in natural spaces and a passion for listening to them.

Some years later, my interest in sound led to undergraduate studies in Music at Oxford Brookes University, where I had the opportunity to take modules in Electronic Music and Electroacoustic Composition. I continued to focus on these areas (and explored elements of soundscape composition) during a master's degree at the University of Manchester, after which I took some time away from academia. During this period, I learnt about a field recording workshop hosted by Chris Watson and Jez Riley French at Orford Ness (UK), which I thought would be a great way to get back into creative composition. After attending the workshop and buying my first portable recorder, it became clear that field recording and studio-based composition were elements I wanted to explore in my creative practice.

At around the same time in 2016, I applied for a PhD at the University of Birmingham, developing the thesis topic 'Blurring the lines between field recording, soundscape composition and acousmatic composition.' As my research progressed over the next few years, I went on multiple field-recording trips to gather source materials for electroacoustic works, exploring natural spaces in countries such as South Africa and Finland. My research at Birmingham was instrumental in terms of finding my voice as a composer; I developed a much greater understanding of wildlife sounds and how to compose with them, carving out my own aesthetic approach in the process.

I'm very interested in working with recordings of natural soundscapes and wild species, and I've developed a particular interest in nocturnal soundscapes following a commission for a piece exploring this theme in 2021 (Audio 6.1). I'm also fascinated by the relationship between environmental sounds and humans, so a lot of my work comments on issues like climate change, light pollution and species extinction.

A few years ago, you recorded Common Hippos along a stretch of the Limpopo Bushveld in South Africa. You used these recordings in two soundscape compositions entitled 'Paddabolela' and 'Mmabolela'. Can you explain a bit about your trip to South Africa and your approaches to gathering wildlife sounds for each piece?

In 2017, I went on the Sonic Mmabolela field recording residency run by Francisco Lopez and Barbara Ellison at Mmabolela Reserve, in the Limpopo province of South Africa. We had a wonderful guide on the trip, William Mosima, who became a close friend and kept me alive when we had to tear through the bush with a machete or encountered a Black Mamba. He and his family shared so much knowledge on where to find different species, and he sat patiently for hours while we all recorded.

The residency focused on creative approaches to working with environmental sound recordings. My personal aims were to gather field recordings to use for my PhD portfolio and to explore new sonic environments; I didn't have particular species or creative works in mind, and I went without any real expectations as to what I would discover. In addition to theoretical discussions, we would have two or three recording sessions per day, each lasting four hours and with plenty of opportunities to leave drop-rigs overnight and collect them the next morning. Since we travelled and recorded as a large group, we would typically set up our drop-rigs in one location, then move to a secondary location where we were able to explore the landscape, make additional recordings, or practice extended listening techniques.

I used a mixture of ambisonic and stereo recording methods in the field. My main set-up was a portable recorder, and a tripod-mounted ambisonic microphone protected by a windshield. I also used a drop-rig consisting of a small handheld recorder and two omnidirectional mics spaced about 40 cm apart, which I often left hanging from a tree in a waterproof bag or placed very discreetly under rocks or foliage. As I gathered sonic materials, I would take notes on what I found most inspiring, with compositional work very much in mind.

The residency culminated in the composition and presentation of a piece by each participant, using original materials gathered in the field. 'Mmabolela' is

Nikki, setting up recording equipment for the Sonic Mmabolela field recording residency in South Africa. An artificial binaural head is pictured in the foreground. (Alex Davies)

actually the final composition in a three-piece series featuring recordings made on the trip. During the residency, I worked in first-order ambisonics to compose and present the first piece, 'Paddabolela' (Audio 6.2). Back in the UK, I spatialised 'Paddabolela' over an eight-channel system and went through a similar process when composing the second piece in the series, 'Limpopo'. For the final piece, 'Mmabolela', I wanted to create a fully immersive, 360-degree listening experience (Audio 6.3): to achieve this, I worked in third-order ambisonics using both ambisonic and stereo field recordings. The raw ambisonic recordings were used to articulate wider soundscapes, with detailed material (gathered using other recording methods in the field, such as AB stereo) processed and encoded to create juxtaposed perspectives in 3D space over 16 channels. Throughout the piece, the listener moves through a series of micro and macro sound worlds, both internal and external, with ambiences and close-up spaces weaving around the audience as part of an immersive sonic experience.

How does your studio practice interact with your recording practice? Do spatial explorations in the studio (especially when working in ambisonics and other multichannel formats) have an impact on how you approach and record species and habitats in the field, and vice versa?

My process is pretty much the same no matter what format I'm working with in the studio: listening; field recording; researching and contemplating; editing; processing; composing; and spatialising. I go about everything in a very intuitive way, letting the sounds guide my practice.

When I'm out in the field, I'm not influenced by studio factors such as loudspeaker layouts or software-based ambisonics work. I have some equipment and techniques that I constantly go back to regardless of how each piece is composed and spatialised, including a mix of ambisonic, stereo and mono recording devices/approaches which can be used to create both stereo and multichannel works. If I know a piece will be performed on a multichannel system, I might try to get more ambisonic recordings on location, but I usually just use the mic that works best based on the sounds I'm hearing. I would say that my studio practice has influenced my recording practice in the sense that I now listen to sounds from both a micro and macro perspective: I try to find panoramic 'beds' of sound along with more detailed, close-up sounds for layering and abstraction. This is more about the compositional approach I favour than anything else.

Working the other way, spatial considerations in the field sometimes inform my studio work. When I'm shaping sounds, I often want them to reflect my direct experiences, which obviously include spatial information. In the final section of my piece 'Nocturnal Wanderings' (composed using field recordings made in Queensland, Australia over a six-month period), I used a base layer of insects to fill a composed 360-degree space, then layered bird calls from curlews, Eastern Whipbirds, kookaburras and other species over this sonic base (Audio 6.4). I then tried to recreate the direction the sounds came from; for example, the Eastern Whipbird material emanates from the front-right of the listener, which represents the direction I was facing in the field when recording the species using a stereo pair of omnidirectional mics.

With sounds that have been abstracted out of their spatial context through processing and so on, I'm more creative with positioning. Speaking more broadly about how studio and field are related for me, I think my interest in spatial audio and ambisonics is less about recording spatial relationships and behaviours, and more about using immersive audio techniques to take the listener on a journey and evoke an emotional response. This may, in turn, change people's perspectives and raise awareness of environmental and biodiversity issues.

You recently travelled to Australia, where you spent lots of time recording around all three UNESCO biosphere reserve sites in Queensland. What were your objectives and key experiences during the trip, and what recording/ compositional workflows did you use?

In 2024, I spent six months on the Natural Ecologies artist residency, which is part of the Refinery programme with the Sunshine Coast Creative Alliance and University of the Sunshine Coast. My research project on the trip, *Visualising*

Nikki shares some work in progress for 'Nocturnal Wanderings', at the end of the Natural Ecologies artist residency. (John Lucy)

Nocturnal Soundscapes, focused on nocturnal wildlife sounds which largely go unnoticed, highlighting the role sound can play in enhancing ecological awareness and monitoring the biodiversity of species which are active at night. In addition to the biosphere reserve sites in Queensland, I went to other areas such as the Daintree Rainforest (where I recorded Spectacled Flying Foxes) and Magnetic Island (where I made some Koala recordings).

As part of the residency, I became a research fellow at The University of the Sunshine Coast and the Creative Ecologies Research Cluster. This meant I had access to new technologies and facilities, including the CAVE2 system (16-channel audio, 320-degree video). Alongside regular mentoring from sound artist and composer Dr Leah Barclay, I took part in field recording trips to K'gari, which is a World Heritage–listed sand island, and went on various boat trips to record Humpback Whales. I met lots of other researchers in the department, got to know conservationists and artists working all over the Sunshine Coast, and was granted out-of-hours access to a number of nature reserves for recording purposes.

What was really incredible during this time was the willingness of other artists, researchers and the general public to get involved in my project and support my research. At the end of the residency, I shared work in progress for my audio-visual piece 'Nocturnal Wanderings' (visuals produced in collaboration with John Lucy); in February 2025, I completed audio for the piece during a residency with the Immersive Audio Network in the UK, using a 25-channel loudspeaker system (Audio 6.5).

The 25-channel system used to finish Audio work for 'Nocturnal Wanderings'. (Nikki Sheth)

There were a couple of sites that I kept going back to whilst on the residency. The first was the Maroochy Wetlands, which contains five different vegetation communities and one of south-east Queensland's largest groups of Mangrove Ferns. The other was Mary Cairncross Scenic Reserve: a protected subtropical rainforest with 141 bird species and 68 mammal, reptile and amphibian species, 8 of which are near-threatened or threatened. The most interesting sounds I recorded at these sites were from amphibians, including species such as the Great Barred Frog (Audio 6.6) and the Striped Marsh Frog (Audio 6.7, 6.8).

As a result of my interest in nocturnal sounds, I've become fascinated by the sounds of bats. Most of their vocalisations occur above the human hearing range (i.e. they are ultrasonic), but in Australia and in other countries, flying fox species produce sounds we can hear without using a bat detector. I became slightly obsessed with flying foxes during the residency, experiencing Spectacled, Little Red and Black Flying Foxes up close and seeing first-hand how education can change perspectives on these species. During a week-long trip to the Daintree Rainforest with my partner and collaborator John Lucy, I managed to record a colony of Spectacled Flying Foxes (Audio 6.9). Shortly after that, I attended an event run by the Sunshine Coast Biosphere where I was introduced to council member Tyron de Kauwe, whose work involves changing negative opinions on bats in the local area

A selection of photos from Nikki's residency in Australia. Clockwise from top left: Nikki, recording in the Daintree Rainforest (John Lucy); Maroochy Wetlands astrophotography (Nikki Sheth & John Lucy); a bat fly-out at dusk (Nikki Sheth); roosting flying foxes (Nikki Sheth).

and farther afield. After witnessing and recording over 3,000 Grey-headed Flying Foxes leaving their roost (Audio 6.10), I was asked to get involved in a project for the tenth annual Australasian Bat Night; this led to the development of an audio-trail work called 'Beyond the Branches' (commissioned by the Sunshine Coast Council and in collaboration with Bernice Wood), which focused on changing negative perspectives on Grey-headed Flying Foxes through education and sound.

There were lots of things that didn't go to plan in Australia. Firstly, I didn't have access to the equipment I normally use; I recorded only in stereo and mono, since I didn't have an ambisonic microphone available (although I was able to encode these sources when working with ambisonics back in the studio). There were challenges with underwater noise pollution when trying to record Humpback Whales, even though we went out after dark to avoid boat and propellor noise. Some situations also got quite dangerous: I encountered crocodile hunters in Daintree and some important equipment failed on a sailboat at one point. In the end, I let go of my expectations and just decided to go with the flow. That's something I've learnt to do with sound recording over the years, and it's paid off.

Can you tell me more about your academic work and how it has shaped your approaches as a recordist and composer?

My doctoral research at the University of Birmingham, and the portfolio of stereo and multichannel works associated with it, drew on environmental field recordings from around the world. I had access to some fantastic electroacoustic studios during my time at Birmingham, and I was free to develop my practice in stereo and multichannel compositional formats. My practice-based research objectives were: to investigate the role of sound recording within genres of field recording, soundscape composition and acousmatic music; to evaluate how sound artists use field recordings in their work; and to identify where my practice lay within the wider disciplines of field recording, composition and sound art. It's difficult to overstate the influence my PhD research has had on the projects I engage with and how I currently approach things.

As an academic, I'm especially interested in cross-disciplinary approaches through working with other researchers, organisations and collaborators: this shapes my approach as a recordist and composer, since I always end up doing some background reading around the species I'm focusing on. I look into scientific or ecological information on their habitats and contact experts and local community members who are passionate about the natural world. My key aims as a practitioner are to raise awareness and engage with issues relating to conservation, biodiversity and ecology through the medium of sound. As the years have progressed, I have found that sound is a very powerful tool for conveying wider messages about the state of our environment and the need to protect it.

Before I went to Australia, I spent a year as an associate artist at Open School East in the UK, which resulted in the creation and exhibition of my work 'Bat

Boxes'. Recordings from this installation were used to create another work entitled 'Chiroptera', which comments on the importance of bats for pollination and features sounds of the Azores Noctule (Audio 6.11). Looking to the future, I see my work on nocturnal sounds as a wider, ongoing research project and I have lots of exciting ideas about how to take it forwards.

Do you have any personal thoughts about wildlife sound recording and the creative practices you are involved in? Is there anything you would like to see change, especially for those starting out?

I would love to see more diverse representation in the fields I work in, especially at higher professional levels.

When I first started field recording and working in this area, I barely knew of any women or people of colour doing the same thing. I am inspired by (and have gained a lot of knowledge from) well-known white-male field recordists and composers, but I always think about how I would probably have progressed in my career a lot faster if I had had a female black, Asian and minority ethnic (BAME) role-model to look up to when I was younger. I would have felt inspired and encouraged: not like I was fighting against the norm, or always going to be at a disadvantage or not taken seriously. That's why I try to be a visible role-model for younger female BAME field recordists and sound artists, and why I try my best to show diversity in the classroom. When I started my PhD, it was the first time I had worked with another female BAME recordist, composer and sound

An infrared photograph of Nikki recording in the Azores. (John Lucy)

artist: Prof Annie Mahtani, my supervisor and mentor. It made a huge difference to my confidence and aspirations to have a successful role-model who was also a BAME female.

I think it's quite difficult to make progress as a female in music technology, where you often end up working in isolation and need to figure things out independently. This area is white-male dominated, so I would love to see more technical guidance and knowledge shared widely through gender-inclusive workshops, masterclasses and events. I would also love to see wider access for female beginners to access equipment, resources and funding.

I often hold field recording and soundscape composition workshops in schools, and teach short courses where I am very open about techniques, equipment and compositional methodologies. I want to share my knowledge and encourage as many people as possible to be creative and feel inspired by sounds of the natural environment, regardless of gender identity, ethnic background or anything else.

Axel Drioli

Adventure never seems to be too far away for Axel Drioli, who recently embarked on two challenging expeditions along the west coast of Africa: Wings Across Continents (WAC) and Sounds Across Continents (SAC). During these trips (the first of which was undertaken with his brother Ario), he traced a long and arduous route along the East Atlantic Flyway in search of migratory bird species, engaging with locals through portable VR and spatial audio technologies and sharing his wildlife sound recording experiences with community members each step of the way. Having followed his progress with interest, I caught up with him during his second expedition to discuss his background in post-production and spatial audio, community engagement work, the logistical and fieldcraft challenges of recording wildlife on the move, and a number of other topics. *MF*

How did you become interested in wildlife sound recording? What inspires you and why do you do what you do?

When I was younger, I had a huge interest in electronic music and did quite a lot of DJing. At the age of 23, I took a one-way ticket to London to build a career as a DJ and play the music that I loved. After three years on this path, I decided to study Sound Engineering at SAE London. I had been curious about the spatial aspects of electronic music as a teenager and was still quite interested in this area, so when someone let me listen to a binaural [virtual barbershop] experience on headphones at university, I thought it was amazing! After that, I started recording music and soundscapes in spatial audio.

After a while, exploring music and sound for cinema didn't make much sense: in these areas, I felt that the frontal sound image was still the main emphasis and the part that people paid most attention to. I decided to focus on field recording, exploring lots of sounds around urban and suburban London (passing cars, public transport, etc.). On one occasion, I recorded a dawn chorus in passing—not really focusing on what I was doing—and was blown away by how engaging it was. I started recording lots of natural soundscapes after that, but I had no idea whatsoever about animals or the sounds they produced. Then I found out about the WSRS and Chris Watson's work. I ended up going to one of Chris's workshops in 2017: he was demonstrating parabolic reflector techniques, and that was really where my interest in wildlife sound recording took off. I began learning as much as I could about bird identification and field recording. Eventually, I formed Sounding Wild (with the goal of creating stories and experiences which draw attention to wildlife sounds in local environments), and launched the two projects I've been doing.

Looking back on how my career has developed, it's clear that my relationship with recording technologies for post-production and spatial audio brought me closer to wildlife and got me out there exploring soundscapes and species. It seems a bit contradictory when you think about what 'technology' and 'wildlife' mean for most people (i.e. how opposed they seem), but it's true: if it hadn't been for audio technologies, I probably wouldn't be on the path I'm on today, exploring the natural world.

Tell me a bit about your first expedition, WAC. Why did you decide to go on it and what were your main objectives with it?

Shortly before the COVID pandemic, I was working on my first documentary project in Cambodia. When I had to go back to the UK at the end of 2019, I realised that if I had been able stay on location for a few more weeks, I could have recorded Asian Elephants. But there was no practical way to do it, because I didn't have the necessary audio gear with me. I thought, 'How cool would it be to have all of my equipment with me all the time, so I don't miss these kinds of opportunities?' When I arrived back in the UK, I decided that I wanted to plan a vehicle-based expedition that would allow me to transport everything I needed for recording. I had heard some things about overlanding and had always wanted to travel, so I came up with the idea of driving through Africa. I discovered that there are avian migratory routes or 'flyways', and that one of these—the East Atlantic Flyway—goes all the way through Europe to southern Africa. That was the moment when things connected together in a big way. I realised that by following this migratory route along Europe and then the western stretch of the African continent, I wouldn't have to focus on one bird: instead, I could follow a range of different species and learn about them during my travels. Most importantly, I could involve local communities in my adventure by creating wildlife documentaries as I went.

After mulling all of this over for a while and realising it would be a bit challenging to accomplish on my own, I called my brother Ario and said, 'Hey, I'm going to do this project! Do you want to come?' And he was up for it! He's a photographer and filmmaker, and as it turned out, he developed a connection to wildlife at roughly the same time as me. We just hadn't really talked about it; it had happened independently, completely naturally. He had almost finished university as well, so it made sense for him to join me once he had completed his studies.

Together, we developed the concept of creating a wildlife documentary for (and about) each key area we visited along the East Atlantic Flyway. Through Ario's area of expertise, photo and videographic elements became a possibility and we realised that we could include my audio recordings within a portable VR experience, using a headset and a 10.2-channel loudspeaker system which could fit in my pickup truck. We worked on this VR idea for one year, all the time making money on another [database] project to allow us to self-fund the trip. Independent funding was very important to us, because we didn't want a major funder to tell us what to do; we wanted to go on our own journey exploring a theme we were both interested in. After all of this preparation, we arrived at a project concept based on the unifying themes of bird migration and community: WAC.

What about preparations for the WAC expedition? Did you do any testing closer to home, and if so, did you receive any feedback from members of the public?

The key thing to sort at the start was obviously the right kind of transport. We bought a pickup truck which we both used for a while. Then we bought an additional four-by-four for my brother, because living in one vehicle together was just too much! We also decided to connect with NGOs at a fairly early stage, contacting the most relevant organisations so they could put us in touch with people who could guide us on our journey. As luck would have it, my car mechanic knew someone at the Wildfowl and Wetlands Trust (WWT); through that connection, we were put in touch with the Migratory Birds for People (MBP) programme. MBP consists of a network of wetland centres across Europe and West Africa, and having contacts in those centres was ideal for us, given the route we were taking and the topic we were exploring. Everyone we got in touch with was incredibly helpful when they heard about what we were trying to do.

We tested the project in Europe before heading to Africa, launching our trip in and around the Urdaibai estuary in the Basque region of Spain. We developed our first documentary there, testing and drafting things to see how well the whole process worked. Then we went to Catalunya and on to the Camargue region in France, doing the same thing. People really liked it, so it gave us a bit of confidence for the upcoming African leg of our journey. For each documentary, we structured our approach into one week of recce and research, two weeks of filming and recording, and one week to make the final production and show it: so one month of work for each documentary experience that we presented.

To make sure the VR experience worked properly, we paid someone to develop the playback software for it. It was very important that the material was easy to edit on the road; we knew that we would be living in pickup trucks, we would have limited battery life and there would be no way to do intensive VR editing on my laptop. We needed something fast and simple which could run on a mid-range computer, so the visual element ended up being slideshow based. The software that was made for us utilised a DAW as a playback engine: every time a marker was reached within the DAW editor, an OSC message would be sent to the program running the graphics and the visual experience would shift to a new slide. Each slide could be a video or an image, so it was a really flexible set-up.

Testing different venues and gauging audience reactions was interesting. In the Basque country, we showed our documentary in a wildlife reserve visitors'

Axel edits Audio from the back of his pickup truck. All equipment and supplies for WAC had to be carefully arranged and packed. Numerous customisations were made to each pickup truck prior to, and during, the expedition. (Ario Drioli)

centre, so it was always going to be well received. In Catalunya, however, we did the whole thing in a campsite: the kind of environment where people and their pets often aren't particularly respectful of surrounding habitats. I remember one person coming to the VR experience we had set up there, when we were showing close-up detail of a Barn Swallow nest (Audio 7.1). At the end, I asked them what their favourite bit was, and of course they mentioned the Barn Swallows. I said, 'You know, they are just over there above the sink in the campsite, where you wash your dishes every day.' And they couldn't believe it! For the next three days we were bumping into each other and this person was explaining how they were watching and listening to the swallows around the campsite.

That was when we realised we were making meaningful connections with people and bringing them closer to other species. Here was someone who had drifted over to explore the VR element because it looked 'cool', but they went away knowing the name of the species that was featured. At that stage and with everything being received so well by the public, we knew we had a workable model for Africa.

Were there any fieldcraft challenges that you had to deal with along the East African Flyway?

The Banc d'Arguin National Park and UNESCO World Heritage Site in Mauritania was one of the most challenging locations for sound recording, especially in terms

A Bar-tailed Godwit flock at Banc d'Arguin National Park, Mauritania. A week of observation was required to establish the best placement times and positions for a series of drop-rigs used by Axel. See Audio 7.2 for the corresponding recording. (Ario Drioli)

of logistics: for example, it took over two hours of desert driving in the area to reach a location where only water, rice and tinned fish were available to buy. It's located where the Sahara Desert meets the Atlantic Ocean, and in winter, millions of waders (Bar-tailed Godwit, Common Ringed Plover, Dunlin and many other species) congregate there.

When we arrived at the Banc d'Arguin, there were huge flocks of waders taking off and they were audible hundreds of metres from where we were standing. It was absolutely incredible. I had to get my equipment close to those flocks, but I wanted to do it respectfully, without disturbance. I knew that the best approach would be to leave some mics and recorders unattended, then retreat to a concealed position to observe; however, this required quite a bit of planning because we were in the middle of the desert and the flocks were moving in conjunction with the tides. I had to place the equipment in areas where the tide came up to—where I knew the birds would be pushed to—and it took us roughly a week of observation to figure out all the different factors: where the flocks would go, the best and highest tide times (the higher the tide, the more the flocks got funnelled into one spot), and so on.

For each rig I left out, I had to mount the mics and recorder on the upper portion of a tripod. The recorders were placed inside dry-bags, and the mics (first- and second-order ambisonic) were placed above the recorders and had furry windshields attached. Each tripod had to be high enough for the equipment to be above water, but not so high as to give the impression of a tree (I didn't want the birds to land on it!) or be obvious to other people, since we weren't sure who else was in the area. This drop-rig method didn't really create any disturbance to the birds, because each one looked like a piece of debris which had beached during tidal flow.

In the end, I had three of these recording rigs on tripods, operating simultaneously. I would put them out at low tide and head back to the shore to observe, retrieving them the next day and downloading the audio files to my laptop. I couldn't use long cables to monitor the sounds because sometimes I'd have to retreat back 300 m or more across the sand. After about a week of trying all of this with very little to show for my efforts, I was on my way to pick up one of the rigs when, from a distance, I could see that the bottom of the microphone windshield looked a bit matted. I knew something had happened. That night, the tide had been much higher than usual and saltwater had gone over the top of the dry-bag. The recorder was just floating there inside the bag, completely destroyed: the only thing that was intact was the internal SD card! Ironically, I got the sounds I wanted on that attempt (Audio 7.2). That's how wildlife recording goes sometimes…

Were there any notable challenges on the African leg of the WAC expedition, in terms of logistics and safety?

Before going to Africa, I was really nervous about all of the logistical aspects: equipment and car insurance, permits for the vehicles, safety, everything. I was freaking out about it. We knew we were going with £40,000 worth of

Axel and Ario's modified pickup trucks, both of which were used throughout WAC. Note the solar panels and essential supplies mounted on the roofs. (Ario Drioli)

equipment—with tourist visas—through Morocco and Mauritania, and we had heard stories about people having things stolen.

Six months before embarking, I started digging for information on what driving through Africa is like. I found social media groups and YouTubers involved in overlanding and read a couple of books. It turns out there are overlanding communities around the world: they have so much information about where to go, what to do when you get stuck, how and where to ask for help, which border crossings to avoid and lots of other things. We would occasionally hang out with these people, some of whom had been doing UK/Africa round trips for over thirty years. We also had to put safety procedures in place before going. I had a list of Italian and British embassies for most countries. We had satellite communicators which could send SOS messages to friends and family. Every night before going to sleep, I would send a message to close family members, two of my best friends and my ex-girlfriend; so if there was a problem, someone would know. In the morning, my brother would do the same thing with family members and a couple of his friends.

There were some scary moments, for sure. In Morocco, two cars ran into the back of my pickup truck. We had one occasion in south Mauritania when we heard gunshots and decided to leave the area to be on the safe side (although it sounded like poaching, to be honest, not automatic gunfire). Apart from these incidents, I never really felt unsafe in Africa. But I was always wary, just as I would be in Europe or anywhere else. I'd look at people and locations carefully, then decide whether it was safe to sleep or stay there.

Tell me a bit about 'Daphne', the customised four-by-four you used for WAC. You created a roof-based rig to enable wildlife recording whilst seated in the vehicle. Can you explain a bit about that and any other modifications you made?

The first modification was done with a bit of help from an independent microphone-maker and friend, who I've been working with since 2017. I decided to customise the roof rack on the pickup truck to hold a quadraphonic array (a 90 cm, 4-channel spaced array of mini omnidirectional mics). After a bit of poking around, I discovered you could remove the plastic caps on the roof rack of the vehicle; this left little plastic squares exposed, which were perfect for adding mics to discreetly. I got some basic omni capsules, soldered the cables to them and routed them through the roof rack to the back of the pickup cabin. After this, I drilled one hole in the cabin and routed the cables to the front storage compartment where I had my portable recorder. I also improvised some wind protection for the mics using tea strainers with furry covers attached; these custom windshields were secured using cable ties, which meant that I was able to cut them off and conceal them when I came to border crossings, so they wouldn't attract unwanted attention! With this set-up, I was able to record fruit bats in the capital of Mauritania, as well as monkeys and some interesting village ambiences: all whilst sitting in the vehicle.

The second modification involved mounting a miniature mic outside the pickup truck, then drilling a couple of holes and running a cable through to a microcomputer inside. I had that little computer running for most of the trip and it made

Axel, driving 'Daphne'. (Ario Drioli)

fairly accurate automatic identifications of all the bird species around me. That was really cool, and the technology behind these kinds of species ID devices is only going to improve over time.

A big focus of WAC was engaging local communities through spatial audio. Can you explain a bit about how you achieved that?

We took the same approach to engagement in Africa as we did in Europe. The whole point was to embrace equality and to build connections; for me, that meant bringing the same experience we showed in Europe to African communities. We wanted our VR documentaries to be shown in villages, universities, at the side of the road, wherever. That was our vision: to bring respect and something inspiring. We also wanted to show that birds don't have any borders. The Common Nightingales (Audio 7.3) and Willow Warblers (Audio 7.4) I recorded in Sierra Leone would have travelled from Europe, like us, and they would go back for summer. And it would be the same for the Eurasian Reed Warblers I recorded in Guinea-Bissau (Audio 7.5). These examples of connecting through other species were inspiring and embodied the message of sharing that we took with us on the road.

In this vein, I would encourage wildlife sound recordists around the world to share with communities. Take headphones with you (or loudspeakers, if you can)

Local engagement was an underpinning element of WAC. Here, community members in Guinea-Bissau experience Axel/Ario's VR documentary, created using wildlife recordings and images from the local area. (Ario Drioli)

Axel, recording individual species using a parabolic reflector. In this case, a monopod is being used to prevent unwanted handling noise travelling through the reflector structure to the microphone. A mini portable recorder has also been mounted directly onto the reflector handle. (Ario Drioli)

and let people listen to what you're doing. If more people did this, it would be a game changer. Many sound recordists gather sounds and bring them back home for use for their own commercial projects. I've done this. But to play these sounds to people, to tell them stories and demonstrate connections with local communities: I feel that's being mindful and giving something back.

We showed the documentaries in different languages, which was really important for us. The last thing we wanted to do was force our own opinions and perspectives on others, because we were very conscious of European colonial impacts on the places we were visiting. This was even more apparent when we witnessed some of the poverty along the way: a lot of that was very hard to take on board, but it's real and we shouldn't shy away from it just because we aren't used to seeing it. I knew that was a part of the journey, and we had to be respectful to people and acknowledge our place in all of it.

You are currently on your second expedition through West Africa, SAC. What are your goals this time?

This one is all about migratory dialects. It's a solo follow-up to the WAC expedition, starting from Sierra Leone and continuing along West Africa on a quest to find

European migratory species. Through listening and recording, I'm exploring whether migratory birds absorb local sounds into their vocabularies. This time, I'll also be recording local stories/music and collaborating with musicians (the immersive audio experience used for WAC will still be available, but I'll be incorporating musical and cultural elements into it).

On this trip, I wanted to find more ways to engage with people, especially younger people, and allow them to have some fun with wildlife sounds. With

An eight-channel array consisting of two quadraphonic configurations organised in a 'cube' arrangement, mounted on a tripod. (Ario Drioli)

this in mind, I've been using birdsong to make some beat-based music which local people can play along with. I've also been making a huge effort to connect with charities. In Sierra Leone, for example, there's a charity called WAYout (of which I am now a trustee) which helps underprivileged people—mostly youths—to engage with society by teaching them artistic skills like music production, filmmaking, podcasting and painting.

The other day, I put together a hip-hop beat in collaboration with some kids involved in WAYout, using a recording of the locally familiar Hadada Ibis as a basis for rhythmic structure. We played it back and all the children were singing along to it. It was amazing to experience, and maybe it will encourage a greater connection to that species and the wildlife here. That's my hope on this journey.

Jana Winderen

Jana Winderen is perhaps best known for her underwater recordings and associated creative projects. In this interview, she discusses her recording techniques through the lens of her artistic practice, tracing a series of exploratory paths from sonic experiences in the field to public presentation in galleries, museums and other spaces. Areas covered include: differences between open-air and underwater recording methodologies; her work in the Arctic, listening to Bearded Seal and Walrus; boat-based hydrophone techniques; the nature of plankton; and recent field recording trips along the River Thames. Around these topics, Jana took time to reflect on her earliest influences and experiences, coming full circle to the wildlife she is currently observing outside her studio in Norway. *MF*

How did you become interested in wildlife sound recording? What inspires you and why do you do what you do?

During my childhood, I spent a lot of time alone in the forest, and out on the sea in my rowing boat. I also spent time outdoors with my grandparents, picking berries, fishing, and learning about rare plants, mushrooms and leeches. When I was around seven years old, I learnt about the environmental impacts humans were having on wildlife and habitats. I became interested in marine biology and studied a lot of mathematics, chemistry, fish ecology and biochemistry at university. I had intended to study for a Marine Biology degree, but eventually gravitated towards Art at both Falmouth University, and Goldsmiths, University of London.

As a creative practitioner, I started recording sounds because I no longer wanted to make or work with physical objects: instead, I wanted to work with material which didn't take up any space. In my second year at Goldsmiths (in 1992), I went out recording along the River Thames, around Greenwich. I started using light, temperature and humidity sensors, and made motion-tracking installations with sound activation. I also created my own contact microphones and made aeolian harps with them, putting them into water and so on. All of this was pure exploration.

After working with sensors for a while and spending lots of time in studios, I decided that I wanted to spend more time outside in search of very small

Icefjord Kangia, Ilulissat, Greenland (2007). During her time here, Jana recorded with hydrophones both under the ice and within the ice itself. (Jana Winderen)

Folgefonna, Norway (2007). Using hydrophones, Jana records meltwater 25 m down in the glacier crevasses. (Åsmund Bakke, Folgefonni Breførarlag)

sounds which humans aren't very good at noticing: things like invertebrates eating leaves, bat and rat vocalisations, insect stridulations, fish and underwater insect activity, and snail sounds. After borrowing some professional hydrophones from an audio supplier in Oslo, I fell in love with how they sounded and started saving money to get a pair of my own. In 2006, I became totally fascinated by the powerful sounds of Vatnajökull in Iceland, as well as the Jostedalsbreen and Folgefonna glaciers in Norway. In 2007, I brought my own hydrophones with me to Greenland to listen to melting glaciers. I also started to listen to fish, including wild cod.

I got very excited about these small and highly detailed underwater sounds that we don't usually pay attention to, and my fascination with underwater sound worlds has fed into all of my artistic work up to the present day.

You are known for your sonic explorations of underwater ecosystems. Can you outline some key differences between conventional (open-air) recording practices and underwater recording techniques? What do you find appealing about listening to underwater habitats and species?

When working above water, the sonic 'colour' of each recording is affected by things like mic construction and positioning, polar pattern, and whether a windshield has been applied. When recording underwater environments, there are just as many sonic colourations to be aware of and a lot of these are the result of hydrophone positioning. In Svalbard, for example, I recently worked with four hydrophones under the ice. Each one was positioned at a different depth (90, 30, 20 and 10 m), and each one provided a very different perspective on the underwater sound environment. I always use my ears to guide hydrophone placement and never make any exact measurements. Just as with open-air recording, it is very important to spend time in a place to gain a sense of the habitat and observe species behaviours over a period of days, weeks and sometimes months and years.

I like to think and learn about how the ability to listen developed in evolutionary terms, from underwater to above water, and how the human body and its structures adapted and changed as a result of being exposed to very different environments. I'm fascinated by creatures like mudskippers and lungfish (which live both above and under water), not to mention all of the insect and beetle species and their different methods of breathing at each

The Silver Bank, Dominican Republic. Jana records underwater sounds with a pair of hydrophones for her project *Silencing of the Reefs*, during the TBA21–Academy residency in 2013. (José Alejandro Alvarez)

developmental stage. With these different breathing methods and perspectives in mind, I often make simultaneous recordings above water, to complement my hydrophone recordings; I sometimes do this in the ultrasonic range, either with a bat detector or with conventional microphones whose frequency range extends beyond 20 kHz.

Over the years, I've found that the most important element in my practice is simply to listen. The compositional process starts out there, in the field. I enjoy sinking into the listening process, absorbing the sounds I can hear and sharing my headphones with other people (boat crews, fellow passengers, etc.). I enjoy seeing the expressions and smiles on people's faces when they listen to something completely new, and these reactions often reveal a kind of childhood curiosity which adults often forget about. I love the raw excitement, the curiosity of hearing something for the first time (which is every time, since you can't hear the same sound again): it sits at the core of what I'm doing and my love of being in—and with—the environment, amongst plants and animals in their natural habitats.

You have worked in the Arctic, recording species such as Bearded Seal and Walrus. Can you tell me a bit about your fieldcraft for these species and what it's like to work in their habitats?

Before considering how to record these species, it is very important to listen to locals and field guides who know the areas you will be working in. This applies to the Arctic as much as for tropical zones, deserts and other locations.

Jana, recording in Svalbard using three identical hydrophones that are positioned at different depths under the ice. (Stuart Trowell)

In places where I've had to work amongst sea ice, I've encountered issues that I hadn't considered asking about, or even thought about. The first time I was in Greenland in 2007, for example, I was recording along the Ilulissat Icefjord, where pieces of ice as big as houses were floating past me; when I considered what I was doing, I realised that if one of the pieces had tipped over, I probably wouldn't have survived because I would have been drowned by the resulting wave. Although it's very beautiful, it's very, very dangerous to work in these areas and you really need to think about what you are doing. Sometimes, you just have to wait for the right recording conditions. When I was growing up in Norway, I experienced severe snowstorms, some of which were so strong that it was a struggle to walk upright. In regions close to and within the Arctic, you quickly learn to respect the weather and understand how it affects everything around you.

When I was recording Bearded Seals for the first time in Svalbard, severe fog drifted in. I remember the urgency of quickly moving back to the ship I was based on; we needed to be able to see Polar Bears clearly, so hanging around in a small boat wasn't a good idea. My first recording of a Bearded Seal in 2016 was really special, because I hadn't heard one before. I was based on a Helmer Hansen Research vessel, and I was using two hydrophones suspended from the edge of a dinghy to record ice floes. My eyes were closed as I listened, because I find that this helps me engage with sounds more. I heard a long, beautiful dropping tone which suddenly stopped. When I opened my eyes to look around, there—no more

Svalbard (2022). Jana is recording with: four hydrophones, positioned at depths of up to 90 m; an ambisonic B-format configuration; a parabolic reflector; three mini omnidirectional microphones; and an ultrasonic microphone. (Andreas Åmli Hansen)

than 30 m in front of me—was the head of a large Bearded Seal. We exchanged looks, then he dived and started singing again. On subsequent recording trips, I've recorded Bearded Seals singing together under the ice, sometimes three at a time and each with a slightly different vocal character. When listening with hydrophones through a hole in the ice in this way, you can hear the acoustics of the underwater space in great detail and gain lots of insight into how the seals interact.

Walrus sounds are extraordinary underwater. They make the most fantastic knocking and tapping sounds, as well as drum-like pulses and otherworldly vocalisations similar to someone flicking a thin metal sheet. They also make lots of grunting and breathing sounds, which are audible through hydrophones. On one occasion, I was listening to these sounds in my usual way (with my eyes closed), enjoying the whole experience; when I opened my eyes, I found that I was looking directly into the face of a huge Walrus! I started smiling. It was incredible to be exchanging looks with her.

Can you describe some of the fieldcraft considerations involved in making underwater recordings from a boat?

It's obviously very important to wear the appropriate clothes for cold/warm environments. As well as proper clothing (and especially when moving over long distances in cold weather), it's important to have some kind of covered or semi-covered boat to protect against rain, snow and wind chill. In terms of protecting my equipment, I always keep things inside transport cases, so I can easily shut the lids and keep my recorder and other equipment safe from water. There are logistical considerations to keep in mind too. On larger scientific research vessels, for example, it can be dangerous to lower a small boat off the side out at sea. You also have to plan well with the crew and scientific researchers, so they understand what you'd like to do.

I find it's often best to be based in one place, with easy access to a small boat: that way, you can wait out any bad weather and visit the same spots over and over again. Calm water is easiest to record in, since there are fewer noises of water slapping off the side of the hull; sometimes, though, these noises might add pleasant sonic context, especially when working from a small wooden boat. I've used spheres attached around my hydrophones to make them more directional, and I've even floated hydrophones away from boats using various methods. I love exploring different approaches to things, and I'm always experimenting and customising my equipment.

Cables are important to think about. I've used different types over the years, exploring various materials, thicknesses and lengths. Some cables can vibrate and resonate in the current, which may or may not be desirable. On the first research vessel I was on in 2006 (in the Barents Sea), I was using unprotected hydrophones on quite thin cables, suspended right off the side of the ship. It was -20 °C, windy and totally dark. Both mics got completely smashed by waves impacting the ship's

Belmar, New Jersey (2022). Jana uses hydrophones to record sounds for her installation *The Art of Listening: Underwater*. (Gavin Browning)

hull! I was working on a BBC documentary exploring cod migration at the time, and I was going to speak directly on-air and describe the sounds I was listening to; but with the damaged hydrophones, this was no longer possible. Since then, I've been very careful about how I deploy my hydrophones.

Sometimes, a boat isn't a suitable option and I end up recording through ice, using a drill (on thick ice) to make a hole and slide hydrophones through. I did this recently around the Barents Sea near Svalbard, when recording Bearded Seals and Walruses. I've also done it many times on frozen lakes. Back in 2006 when I was recording for *Energy Field* (Audio 8.1), I even put hydrophones directly inside chunks of ice to pick up the popping and melting sounds of glaciers in Iceland, Greenland and Norway.

Plankton have been notable subjects of interest for you. What recording methods do you use to investigate their behaviours and habitats, and what listening approaches do you employ?

Talking about plankton is such a vast topic, because there are many thousands of species known to science and probably many more unknown. Some zooplankton make sounds via tail flips, such as *Meganyctiphanes norvegica* [Northern Krill] and *Calanus* spp., although I'm not sure if this is intentional/communicative or just the audible result of locomotion through water. Zooplankton make the largest vertical migration on the planet every day through the ocean.

When I was working on my composition 'The Wanderer', I was thinking particularly about *Calanus finmarchicus*, a very important species in the Norwegian Sea. This copepod is essential for the food web, being an important food source for cod larvae, herring, pollock and other fish, as well as the basic food source for the Blue Whale. In the studio, I was thinking about the various sonic environments plankton drift through as they are carried by the current. I ended up calling the piece 'The Wanderer', since the word 'plankton' (from the Greek *planktos*) means 'wanderer' or 'drifter'. The whole piece was put together from hydrophone recordings I had made in the ocean, in areas where plankton are active.

It's amazing to contemplate this large-scale vertical migration through the water, but I don't feel the need to isolate individual species within it. As a recordist and artist, I am mainly interested in recording wider environments and the relationships between different species within their natural habitats (as opposed to focusing on individual species, which many wildlife sound recordists do). Nonetheless, I have learnt to recognise particular species by listening and observing, and by talking to locals and scientists in the environments I work in.

The projects *Spring Bloom in the Marginal Ice Zone* and *Listening Through the Dead Zones* are also about plankton. *Spring Bloom in the Marginal Ice Zone* (Audio 8.2) is about the very important plankton bloom in the ever-changing zone between the open sea and sea ice. Many phytoplankton get eaten by zooplankton in the marginal zone, and these creatures, in turn, get eaten by fish and mammals. Collectively, the trillions of decaying animal remains, faecal particles and so on which result from this process are known as 'marine snow'. This cloud of particles slowly sinks to the sea floor, becoming food for bottom dwellers in the benthic zone and creating a very important carbon sink in the Barents Sea every year.

With the *Listening Through the Dead Zones* installation (Audio 8.3), I was trying to draw attention to the dead zones in the Baltic Sea and coastal environments that are suffering from hypoxia. This is where eutrophication depletes the water of oxygen and makes it uninhabitable for fish and other creatures, which depend on a certain amount of dissolved oxygen in the water. The amount of dissolved oxygen needed to survive varies from creature to creature: bottom feeders, crabs, oysters and worms need minimal amounts (1–6 mg/L), while shallow-water fish need higher levels (4–15 mg/L). In the Baltic Sea, the threshold for labelling an area as a dead zone is 2 mg/L.

You recently recorded sounds along the River Thames in London, for an installation entitled *The River* (produced in collaboration with Tony Myatt for the Natural History Museum). I'm curious about how you approached the Thames and its species as a collective recording subject. Can you tell me a bit about the project?

When developing ideas for *The River*, I made some test recordings around Embankment Pier in central London. I could hear trains, voices, ferries and the creaking sounds of the pier itself through my hydrophones. From an early stage and because of these experiences, it became very important to me that all recordings for the project were made underwater: people would then understand how easily land-based sounds propagate through water, and how they would be audible to species living there (depending on their sensitivity to sound). During one recording session along the Thames, a dolphin and its young were spotted moving towards Richmond. I remember thinking how loud it must have been for the travelling cetacean family, and how important it was to raise awareness about the impact human activity is having on underwater sound environments.

I decided that my composition would start at the source of the Thames, following its course all the way to the Thames Estuary where the river meets the North Sea. At the source, I recorded lots of underwater insects and beetles making sounds, right at the point where water was starting to flow and gas bubbles were emerging out of the mud. Farther down the river around Cricklade, there was an incredibly rich and beautiful sonic environment of underwater insects and small fish flipping their bodies, which sounded very loud through my hydrophones. As I moved towards London, the human presence naturally increased; I was still able to record, but it was quite challenging. As the estuary opened up, the sonic image widened: sounds from crustacea and the open sea became audible alongside cargo ships and industrial activity on the outer areas of the estuary, looking towards the Maunsell Forts and the London Array wind turbines.

Working in a city environment is very different to working in remote locations. You need to consider safety in a different way. When recording from portions of land or public piers, you might need permission from the authorities. It is also harder to let yourself relax into listening and close your eyes in these environments, since you have to keep an eye (and an ear) open for other humans. In less populated areas, I find that I can relax into the environment and flow with it. Perhaps for a future project like this, I would rent a riverboat and live on the water whilst working. Getting to know all the inhabitants and the species around me over a whole year would be brilliant.

Are you working on any other projects at the moment? If so, are they informed by your established practice or have you taken a different approach with them?

Over the last four years, I have been following my wildlife neighbours very closely, right beside my studio in Norway. I feel very close to the mothers: a deer mum, who this year had three calves; a badger mum, who recently had three cubs; and a squirrel mum. Last summer, there were ten Common Pheasant chicks running around here, not to mention all the Eurasian Blue Tits, Goldfinches, Hawfinches, Greenfinches and Hooded Crows I am sharing the neighbourhood with.

As I have said in so many talks I've given, you do not have to travel to the likes of the North Pole or the Caribbean to learn about—and be amongst—wildlife.

It is all around you, in your back garden, in parks and in other spaces. I recently returned from Brussels where they have rewilded lots of public parks, and it was fantastic to see the biodiversity and liveliness there; I hope this will be done in every city, and that habitats for all species are considered and respected.

When I used to live in Oslo with my dog Charlie, we often walked in a nearby forest where I observed tadpoles, dung beetles, shrews, foxes, and (on one occasion) what I believe was the track of a Grey Wolf. My main interest has always been our fellow creatures and their habitats, so I find it a bit hard to single out particular projects; my whole life has been about observing and learning about our environment, which we share with other creatures and plants.

How do you view your practice in general, and its place within (or beyond) field recording? We have focused on 'wildlife sound recording' during this interview, but as a practitioner, do you feel that your work extends beyond definitions such as 'wildlife' or 'field'?

Although I'm being interviewed about wildlife and field recording subjects here, I have not defined my work as either.

As I mentioned at the beginning, when I started recording sound it was because I did not want to make more physical objects: I wanted to work with material which didn't take up any space (or perhaps more accurately, I wanted to give people an

By positioning microphones immediately next to subjects in the field, it is possible to record extremely delicate sounds, such as this caterpillar eating a leaf. (Jana Winderen)

A spider senses vibrations through its web in Maridalen, Norway. (Jana Winderen)

experience to carry home in their memories). It's all about curiosity and love for life. I have always known what I'm doing, but I have never really labelled it as a particular craft, like 'field recording'.

I realise that field recording has become a thing, so to speak, and that people call me a 'field recordist' or 'sound artist'. I would say that I'm first and foremost an artist who feels at home outdoors, and someone who has a deep curiosity about all the other creatures we share our environment with.

Ángel Pérez Grandi

Ángel Pérez Grandi is someone who is naturally drawn to storytelling through sound. Having carved out a successful career in the world of post-production, he has a firm understanding of the power of audio to deliver important conservation messages both independently, and in combination with visual media. In this interview, he offers an insightful glimpse into his studio work and his interwoven wildlife sound recording practice, focusing on community-based documentary storytelling and the creative potential of natural sounds. He also talks about the importance of weighing career aspirations alongside personal passions and hobbies, ethical considerations for wildlife, and the benefits and joys of introducing children to nature. *MF*

How did you become interested in wildlife sound recording? What inspires you and why do you do what you do?

I love the raw qualities of sound: how it affects us physically and emotionally and how it fires the imagination. I grew up in a family environment in Spain with

Ángel, recording from a camouflaged tent hide. Such hides can be used as low-profile observation positions when recording wildlife at a distance. (Ángel Pérez Grandi)

plenty of opportunities to explore outdoor settings and watch the occasional David Attenborough documentary on television, so the two main seeds of my career were planted at an early stage.

It was only later that my love of nature and sound came together in quite a big way. As a young adult, I had already decided to pursue a career in audio, but I couldn't play any instruments. I ended up studying for a Sound Engineering degree in Bristol (UK). After university, when I had progressed down the professional audio route for some years, I noticed that I was missing a deeper connection to the natural world; so I started to focus more on recording wild species and habitats.

I'm drawn to the stories behind natural sounds: I love the sense of quest and adventure underpinning each recording trip outdoors, and the ever-present feeling of not knowing what I'm going to find. On a practical level, I'm constantly building my personal sound library to use for films and creative projects, but I'm also inspired by the idea of contributing to research through citizen science projects, for example, and collecting and preserving acoustic records for future generations. All of these elements have become interwoven in my practice.

More generally and on a deeper emotional level, recording wildlife has become my pathway to a healthier lifestyle. When I stop and really listen—which recording compels me to do—I feel present, connected and at peace.

Can you give me a sense of how living and working in different locations (e.g. Spain, Northern Ireland, Scotland) has informed your listening and recording practice? Was there a particular moment in any of these places when you felt compelled to record wildlife?

My birthplace, Santander, is a coastal city. A huge portion of the outdoor time I spent as a child there was at the beach. Despite leaving Santander when I was young, I have maintained a strong connection to the sea (and its sounds) as a result of living there.

When I turned 30, I became a dad and ended up in rural Ireland, both Northern and southern, for the best part of ten years. The roar of the ocean and the cacophony of my home city faded out as my ears gradually tuned into this new environment. Of course, I soon realised there was plenty of cacophony in the countryside and that it was not exactly the romanticised 'bird-chirping paradise' I had envisaged; in fact, I realised that human-made noises were much more noticeable and jarring in the countryside than in urban areas. Despite this, the part of County Fermanagh where I settled during this period is still one of the least populated regions on the island of Ireland. Living there with my family presented a huge opportunity to seek out natural sounds and explore the landscape. After a number of cathartic dawn-chorus recording sessions there, it occurred to me that there was incredible detail within the chorus: individual birds were vocalising and they could be recorded separately. So, I investigated…

Another point of inflexion was a series of work trips to the Isles of Lewis and Harris, in the Outer Hebrides of Scotland in 2017. There, I collaborated with artist Julie Brook in some of the remotest places I have ever been to. The areas we lived in were off the grid, with no mobile phone coverage. The work we were doing was not centred on wildlife sounds, but the natural soundscapes of the islands were impossible to ignore. The remoteness of the experience really made me engage with landscape; in the process, the sounds of the elements, as well as surrounding fauna and flora, came into sharp focus and inspired me to explore further.

You participated in an EU-funded project which involved exploring natural sounds within your water catchment area in Northern Ireland. Were there any species that you wanted to document for this project? If so, how did you approach and record them?

In 2021, most of my freelance work opportunities disappeared as a result of the COVID-19 pandemic, so I decided to embark on a year-long project to showcase the natural sounds within my water catchment area in Fermanagh, where I was living with my family. I wanted to highlight the importance of good water quality and how it connects all living things, using sound as a primary storytelling medium.

Common Frogs were one of the first species I recorded. In early March, I visited a pond I had spotted on a previous hike and immediately noticed that the frogs' calls faded out as soon as they detected my approach. From then on, I kept my intrusions to a minimum, preparing an ORTF rig and a pair of hydrophones to attempt a simultaneous above-water and underwater recording. Back in the studio, I realised that these recordings could not be easily layered: the pleasantly balanced above-water soundscape (Audio 9.1) felt completely separate from the underwater scene. In addition, the hydrophones had recorded sources closest to each capsule, so my 'stereo' underwater scene sounded more like dual mono (Audio 9.2). All of these sounds were gathered early in the morning to avoid disturbance from other people. On a subsequent daytime visit and out of pure curiosity, I left a portable handheld recorder out to check if there were any differences in frog-calling behaviour due to human activity; in the process, I managed to record dog-walkers arriving and leaving (Audio 9.3), which added some additional context to the project!

Common Starlings were another subject of interest. Each spring, a pair would nest in the eaves of our family home, just above the children's room. I managed to access the eave cavity through the loft, using a small omnidirectional microphone taped to the end of a boom pole to get some close-up perspectives of their nest activity and species mimicry without any disturbance (Audio 9.4). Based on my own analysis, I think they may also have been mimicking interior household

A mid-side rig (left of image), used to record Eurasian Curlews in Fermanagh. The mics have been placed within a basket windshield, which has been camouflaged with scrim netting. Note how the whole rig has been positioned within a cluster of trees, to blend in with the surroundings. (Ángel Pérez Grandi)

activity, including the rather loud screams of one of my daughters (Audio 9.5)! A recent move to England has allowed me to record huge gatherings of starlings at Shapwick Heath in Somerset; these flocks contain many thousands of birds and are some of the finest nature spectacles in the UK (Audio 9.6).

I also managed to make recordings of Eurasian Curlew, under full licence and in collaboration with The Royal Society for the Protection of Birds (RSPB) Northern Ireland, as well as a local ecologist. Although I consider the recordings to be of average quality, I'm very fond of them: not just because of their inherent conservation value, but also the community connections I made as part of my search for the species. After observing and preparing extensively in early spring 2022, I gained access to a shore which I thought offered the best chance of making a good curlew recording. I had a mid-side rig and a modified parabolic reflector with me, and I based myself in a camouflaged tent hide where I could observe everything in detail. I trained my mid-side rig and reflector on two separate boulders marked with droppings. On the second day, a curlew flew right onto one of the boulders, calling as it landed (Audio 9.7). I got other decent recordings of curlew (Audio 9.8) and later used a similar approach (placing a microphone in the middle of a field and running long cables back to my camouflaged tent) to document the display activity of Common Snipe in the area (Audio 9.9).

A three-channel parabolic reflector configuration, also used to record Eurasian Curlews in Fermanagh. In addition to an internally mounted single mic, two omni mics have been attached each side of the parabolic dish to record the surrounding ambience. (Ángel Pérez Grandi)

To assist with positioning a stereo pair of hydrophones next to stridulating water-insects in Fermanagh, Ángel used a fully extended boom pole. A foam baffle was also used to keep the hydrophones close together, to improve stereo imaging when recording small subjects such as water beetles. (Ángel Pérez Grandi)

Since the project was fundamentally about water, I also recorded sounds of underwater stridulations (Audio 9.10) and other aquatic behaviours using hydrophones. I used a mini underwater camera to scope out species' habitats and assist with ID, which was very challenging and yielded mixed results. The hydrophones were almost always used in a spaced stereo configuration, with a small baffle separating each one: when recording such small subjects so close to the capsules, this gave more pleasing results than a widely spaced pair (something I had learnt from my experiences with Common Frogs). I also tried—and failed—to record the sounds of Northern Pike and White-clawed Crayfish.

Following on from the previous question, did your time in Fermanagh leave you with any significant realisations about recording wildlife?

One of my main realisations was that it takes time, patience and diligence. It's a slow process of informed trial and error. Nature doesn't provide opportunities on demand, but that's precisely what makes the experience of wildlife sound recording so enriching for me: the whole process involves moving at rhythms diametrically opposed to how most of us live our lives in the current 'attention economy'. Wildlife sound recording forces me to think outside the box; it has taught me to accept what comes my way and appreciate it for what it is.

I also discovered that recording wildlife can be quite overwhelming, especially if you have an agenda like a personal species-recording goal, or work commitments (e.g. sound design) which involve gathering sound effects for games, films and so on. In this regard, I would encourage anyone seeking a career in field recording and related areas to consider the line between work and passions/hobbies carefully. What does a healthy balance mean for you? Is it worth turning your hobby as a recordist or sound designer into your main source of income, or is it best to separate work and passion so that there are clear boundaries and one does not spoil the other?

I have also had family considerations alongside my career. In my twenties, I was lucky enough to travel widely; in my thirties, most of my efforts revolved around family life and raising my children in Fermanagh. During this period, I developed a longing for the exploration of remote places, which I could no longer fulfil. However, looking after a young family in Fermanagh instilled the importance of recording wildlife sounds locally and helped me to reconnect with nature through very rewarding activities with my kids. I also started to explore the positive storytelling opportunities of my surroundings, recording in the area and connecting with communities there. Recording locally contributed to an appreciation of place: I began building emotional connections to the place our family called home, and I started caring deeply about local environmental issues. Looking back, I realise that this is how I became interested in nature conservation, and how I ended

Ángel's children, observing wildlife from a hide. (Ángel Pérez Grandi)

up putting so much emphasis on the importance of local community work and storytelling activities as a recordist.

Introducing children to nature is like opening a door to a world of wonder and awe: one that we, as adults, often forget about. Children connect with nature on a visceral level; they don't need explanations. They feel the wind, smell the earth and listen to the birds simply because these elements are there to be experienced. Through exploring the natural world with my own children, it's no exaggeration to say that I have rediscovered nature. When I was outdoors with them, my first impulse (perhaps this is true for most parents) was to share my own childhood memories, somehow trying to relive them through my own kids' experiences. Soon, however, this impulse shifted, and what started as rediscovery through them became pure discovery with them.

I recently listened to 'Echoes from the Barn': an award-winning documentary piece of yours, featuring Barn Swallows in a rural outbuilding. Can you tell me about your recording methods for the piece, and how you adapted your approaches around the human-made structures Barn Swallows typically utilise for nesting?

The last house we settled in in Fermanagh was an old rural cottage, which had a couple of barns nearby. The previous owner was long gone before we moved in, but the hay on the floor of the barns told a story, as did the muddy remains of swallows' nests along the wooden rafters. From the beginning, I could see that there was a lot of potential for a narrative-driven documentary piece focusing on the nesting activity of the swallows (Audio 9.11).

Barn Swallows have evolved to nest in structures; as a result, they have taken advantage of houses, barns and other buildings humans have constructed. They make so-called adherent nests, which they construct by layering mud, grass and their own saliva to bind it all together. The birds return to the area they were born in to breed each year, and their arrival around mid-April or May was always a cause for celebration in our home: they would perch on the electricity wires in front of the house, and their flight calls would fill the sky around our dwellings as they foraged and inspected the area for nesting sites. To help with this process, we would leave the doors of the barns open from early summer until October, and we would make sure there were no access points which would allow our cat Charlie to climb up the rafters.

From a wildlife sound recording perspective, this was an opportunity served on a plate. I knew that the nesting/breeding activity I wanted to document would quite literally unfold on my doorstep. The barn I chose to work with was essentially a controlled environment in terms of human disturbance and activity, which meant that I had the luxury of time to plan and rig my equipment. My initial goal was to record the breeding process from a close-up nest perspective, so my main considerations revolved around negating disturbance and deploying recording gear (microphones, cables, etc.) for an extended period. I also wanted to encompass the acoustics of the barn space, since it formed a part of the species' natural environment and represented an important part of its story.

I rigged the rafters with two mini omnidirectional microphones, trying to figure out which pre-built nests were likely to be reused based on last year's observations, as well as the state of each nest and its position in the barn. All of the rigging was done as soon as I noticed the swallows' arrival in the surrounding area, well before nesting actually started. I laid the cables around the barn interior and placed my microphones as closely as possible to two promising-looking nests without touching or handling them, using the structure of the barn to conceal them. I also clamped a mid-side rig to a nearby wooden beam, pointing it towards the nest and making sure it covered a parallel beam where I knew the adult birds liked to perch. Finally, I constructed a makeshift recording booth at the back of the barn out of cardboard boxes: somewhere I could run electricity to and move in and out of without the birds seeing me.

I made several recordings at different intervals and times of day over the course of two months, in an attempt to document the whole nesting process, from incubation to fledging. There was a lot of quiet time, since the female mostly sat on the eggs as the male went back and forth. After the eggs hatched, things became much more animated and with each passing day, the nestlings started making audible sounds which developed into proper begging calls. Early mornings seemed

A mid-side rig was clamped to a wooden beam and pointed towards the Barn Swallow nest discussed in the interview. A separate beam used by the birds as a perch (visible towards the bottom of the image) was also covered by the configuration. (Ángel Pérez Grandi)

Mini omnidirectional mics were rigged in advance of the swallows' breeding activity, to prevent disturbance. Both mics' cables were neatly routed and taped to the underside of the barn's corrugated roof. (Ángel Pérez Grandi)

to involve more continuous chatter from the adults and their incessant work to keep the young fed was something to behold. At one point, I even recorded the final moments of an insect as it buzzed its last buzz within the closed beak of an adult!

You work extensively in post-production. A lot of readers will be curious about how this informs your approach to wildlife sound recording, and vice versa. What are your fieldcraft considerations when recording wildlife sounds for post-production, and how do the two main strands of your career (recording and studio work) interact?

Wildlife sound recordings are great for setting scenes, painting a mood or establishing a location. Some sound inclusions can be ecologically incorrect, and whilst this may annoy many naturalists and bioacousticians, they should arguably be taken with a pinch of salt; after all, filmmakers are telling a story. Sometimes, a certain sound needs to be used to cut through the overall mix, or to enhance visual material which would not be accompanied by an audible sound in the real world. We should also bear in mind that humanity's disconnection from the natural world is palpable: if post-production experts can keep people engaged through the use of wildlife sounds (even if things aren't 100% accurate), then we can also convey the important message that there is a real world out there to be experienced, and build bridges from the living room to the field, so to speak.

In terms of ethics, there is a tendency in sound effects recording for film (which is my main working background) to try to get a particular sound no matter what it takes. I used to be a part of this approach, but when I started recording wildlife I disengaged from it completely. I realised that sound recording is not the most important thing: respect for the subject is. It is very much like knocking on a neighbour's door to ask for a favour: you would not simply barge in and grab something. With wildlife sound recording, respect comes first and it's essential to consider your intrusions and the level of disturbance you are causing to other species.

When recording wildlife sounds for post-production, there are essentially two approaches I follow: isolation of species/phenomena, or gathering of diffuse sounds. I'm typically either trying to get a clean recording of a specific subject (free from as many background sounds as possible), or an enveloping atmospheric track with a much wider sonic focus. Individual species sounds can be used as they are, or as source materials for designing other sounds and even building entire soundscapes. Atmosphere tracks, which are usually recorded in multichannel formats, are great for creating backgrounds or 'beds' on which other sounds can sit. If I'm recording for a specific project, I'll aim to cover sounds that best match the brief; if a film is location centred, I'll always research the area, figuring out what wildlife is found there, what the habitats and weather conditions are like, and so on. All of these elements inform a list of target sounds. At the end of the whole process, I'll ideally end up with a broad palette of recordings to draw from when I start crafting a soundtrack.

A picturesque view across typical Eurasian Curlew habitat in Northern Ireland. (Ángel Pérez Grandi)

There are several commercial sound libraries available for downloading post-production materials, but they tend to be generic and the sounds often don't fit well with a particular project. Because of this, in the last ten years or so and with the ever-lowering costs of audio technologies, there has been an explosion in boutique/niche sound library production. Even with these new resources, it can still be very hit-and-miss in terms of finding exactly what you need for a project, especially when it comes to locations and definitely when it comes to individual species. Although niche libraries are useful for covering certain elements of a post-production project, I find that nothing beats location-based sound gathering: the recordings simply fit better on-screen.

For both my field recording and studio work, I find it very important to connect with like-minded groups and practitioners. Wildlife sound recording is almost always a lone-wolf activity and freelancing in audio post-production can also be very isolating. Over the years, I've learnt to connect with others through a shared passion for nature, and I've found the overwhelming majority of wildlife sound recording community members to be kind and generous with their knowledge.

Matt Mikkelsen

Matt Mikkelsen has a particular interest in how species exist alongside each other and interact in the natural world. Having recorded in almost every US state at the time of this interview in late 2024, I felt that he was in an ideal position to offer insights into his listening experiences around his American homeland and talk about some of the key challenges he has encountered there. We spoke about everything from military training exercises and how to plan around their sonic intrusions, to post-production work and logistical considerations for long-term recording trips. Throughout the interview, I was struck by his enthusiasm for the natural environments he documents, his deep respect for Indigenous communities, and his ready acknowledgement of those who have helped him on his journey as a sound recordist. *MF*

How did you become interested in wildlife sound recording? What inspires you and why do you do what you do?

I started as a musician, learning how to record music and coordinate live sound events: that was my first foray into audio. Eventually, I decided to go to university

Matt and his mentor Gordon Hempton, standing next to their artificial binaural recording heads. Note the foam wraps placed around each head, for basic wind protection. (Palmer Morse)

to study audio engineering and sound design for film and television. At some point on that journey, I realised that I wanted to spend more time outside and less time in the studio; I'd always loved being and working outside, and I had a real love for nature and the environment. That's when I heard about Gordon Hempton and his work. Long story short, Gordon and I became good friends and close collaborators when I was still quite young, and I got into nature sound recording through his mentorship. I've since established my own place [as a sound recordist], but I still work with Gordon a lot.

I do what I do because for me, listening is one of the most magical ways to exist in—and to try to understand—the natural world.

How do you approach and document wildlife in the field? Do you take a broad perspective encompassing wider habitats and ecosystems, or do you focus on individual species?

I'm much less interested in isolated events (i.e. individual species) and much more interested in how species exist alongside each other and interact in a space. I find that this holistic approach instils an awareness of wider ecological issues: how much noise pollution is in an area, the presence of invasive wildlife, threats to native animals, and so on.

One of my favourite birds, Swainson's Thrush, has this complex, upwardly spiralling song which is beautiful on its own. But for me, when that song echoes throughout a landscape, it tells me a huge amount about the place that species lives in and what its life is like, and that's what I find truly fascinating. Another example would be Common Loons (Audio 10.1). They have some amazing calls. When you hear their sounds reverberate for 10 or 15 seconds across a lake, it's one of the most incredible experiences and it provides so much context about how far their vocalisations travel naturally. I've been asked on several occasions if I've added reverb to my loon recordings. I haven't: I've just documented how their calls propagate naturally through the environment.

Bearing in mind this holistic approach, can you give me an overview of the recording methods you use in the field?

Like most sound recordists, I have a microphone locker that's probably a bit too big!

My early training was mostly binaural. Anyone who tries the binaural approach recognises very quickly that even minor adjustments to microphone placement can have major impacts on how the soundscape is rendered. One of the goals of binaural work is to create a sense of realism: to try to convey sounds as we would hear them naturally, through our own ears. At some point, I realised that the audio industry (particularly sound recording and sound design) was starting to progress more and more into VR and spatial playback formats. Binaural material certainly has its place in that world, but I thought to myself, 'Okay, if I'm going

Matt records massive Pacific coast waves on Rialto Beach, Olympic National Park. A faux-fur cover has been fitted over the ORTF-3D basket windshield for additional wind protection. (Nick McMahon)

to go and catalogue soundscapes all over the world, I want to do it in a future-proof way.'

I tested every configuration I could get my hands on. I think there were about ten of them, including ambisonic, double mid-side and binaural rigs. And what I ended up going for was an ORTF-3D system, which is an eight-microphone array. It offers a lot of flexibility when it comes to post-production: you can just grab a stereo pair from that rig and it will sound wonderful as a regular ORTF recording. You can also use the full eight-channel recording for a Dolby Atmos mix or a VR mix using different playback formats, and the beauty is that it can transition between those formats really easily. Choosing the ORTF-3D method was about trying to record things in a way that was marketable and desirable for people who wanted to use my sounds.

Even though I specialise in ORTF-3D, as a general rule I'm pretty kit agnostic and I bring all kinds of stuff with me. I love using inexpensive drop-rigs incorporating mini omnidirectional microphones, because I can clip them to leaves or twigs, hook them up to a cheap recorder and leave them outdoors for days in all kinds of weather. I would never do that with my ORTF-3D system! I also use a SASS-type rig. I still get a lot of joy from working in stereo (I actually prefer it to other formats), but in my view, the flexibility ORTF-3D offers is unparalleled.

You are heavily involved in documentary filmmaking and regularly collaborate with videographers. Are your recording methods and fieldcraft influenced by working alongside visually focused practitioners in the field? How do you work with (and around) each other?

While I'm primarily a recordist and sound designer, I also do a lot of producing and directing for nature- and conservation-focused productions. I co-founded Spruce Tone Films with [videographer and photographer] Palmer Morse, and my work with him has given me a lot of insight into the requirements of filmmaking and the role that sound plays as a storytelling tool, in conjunction with visuals.

Experienced filmmakers understand the importance of sound. They know that you can have the best visual quality in the world with $120,000 worth of cameras and lenses, but if you have poor-quality sound, the whole production will suffer. Most of the time, the folks I work with understand that I'm thinking about sound throughout the production process, and they give me space to explore that and throw ideas out there as I go. The great thing with Spruce Tone is that I'm the one doing pre-production planning, production sound recording and sound design, so I have a full overview of what's happening from start to finish, including what we're trying to achieve story-wise and how I need to incorporate my recordings.

One of the biggest challenges in my work is understanding the relationship between abstraction and representation. Sometimes it's difficult to overcome the mindset of realism—of direct documentation—as a nature sound recordist. I understand in great detail what species and habitats sound like because I've spent so much time immersed in that world, listening; as a result, when I see visual material reflecting a certain natural environment or situation, my first instinct is often to have something that's accurate and realistic. And sometimes that's not the best thing for storytelling.

There's a fine line between fiction and direct representation of a subject, and it's a balance that we all have to navigate as recordists, especially those of us who are also involved in post-production. As a sound designer (and especially in wildlife and conservation filmmaking), you want to harness emotion effectively, but at the same time there's a need to remain truthful to the experience, to convey a reality. In a lot of nature documentaries, we see things like lions battling each other, wolves tearing apart elk, and so on: these big, dramatic moments. But when you go to a national park, that's probably not going to be what you experience on a day-to-day basis. There's a certain amount of storytelling and fiction with what we do for film, and I think that's okay, because it still reflects what happens out there.

Some of your projects have focused on Indigenous communities in the USA. Have these communities' views on (and intimate knowledge of) wildlife and habitats influenced your own work? Have they made you reassess your own practice as a recordist in any way?

Very much so.

I want to be absolutely clear that I'm not in any way trying to speak on behalf of Indigenous communities: they have their own views, thoughts and voices. Speaking purely from my own perspective, I've observed that Indigenous people in the United States don't really regard themselves as separate from their environment. They have deep and incredibly respectful relationships with wildlife and ecosystems. Even the words that they use to describe things in our environment have made me pause and reflect on my own impact on our planet, and I think that is something that we should pay attention to when we think about things like climate change and biodiversity.

Many wildlife sound recordists are active conservationists and environmentalists: we do this work because we care about our ecosystems. But I think we should remember that conservationists have a problematic history of arriving in places and saying, 'This needs to be protected for whatever reason(s).' Although this may be well intended, it has historically taken place with little or no regard for native peoples. I think it's important that when we enter a place—even if it's not technically a reservation or it's not actively held native land—we need to do our best to recognise the history of those places, environmentally, politically and culturally. And we need to be conscious of the terrible impacts of erasure and colonisation which are still happening not just in the USA, but around the world.

I think that as recordists and human beings, the more meaningful our interaction with the land becomes, the more we can work to help heal relationships and build towards a better planet. That's something that working with Indigenous communities has reinforced for me.

You have recorded in almost every US state. Can you comment on some of the recording challenges you have encountered in the USA and how you have dealt with them?

One of the big advantages of recording in the United States is that we have a lot of beautiful and expansive natural areas: wilderness zones, national parks, state parks, and plenty more. There's a lot of space, especially once you head west of the Mississippi River where you're able to find swathes of uninhabited land. We have everything from arid, high-altitude deserts containing almost no plant life (and very little animal life) to lush temperate rainforests and everything in between. We have parks like Yellowstone and Grand Teton with Coyotes, elk and other mammals roaming around (Audio 10.2, 10.3), wetland habitats filled with birds and amphibians (Audio 10.4), mountain ranges, old-growth deciduous forests, and prairies (Audio 10.5). As you might expect, it's difficult to access some of these spaces: you need four-by-four vehicles, maps, navigation skills. You're not going to have decent phone service. In short, you pretty much have to fend for yourself and I've had to develop wilderness skills in parallel with my recording craft as a result.

It's important for me to research what opportunities there might be sonically, but I also need to understand potential threats to my safety and wellbeing in the USA. There are the obvious dangers of animals like bears, cougars and snakes. Even plants can pose threats: as recordists, we often crawl around in the dirt and navigate through vegetation, but you can hurt yourself doing that if you don't know what you're navigating through. Just being alert and aware is also really important: sometimes, you become so focused on the craft of recording that you lose track of your surroundings and stop paying attention to what's going on around you. You still have to understand what direction you're walking in, whether it's private land, what the different landmarks or geographical features are, whether you're eating enough food and drinking enough water, and so on. Things like that are crucial to keep track of.

Unsurprisingly, there is a lot of military activity happening on public land in the United States. The armed forces have certain pieces of land that they own to conduct training exercises on, but a large amount of training and testing actually happens over public land. That's a difficult thing to navigate in terms of recording wildlife, because you don't really know where they [the military] are going to be and when they're going to be doing those things.

A good example of having to navigate this issue would be Organ Pipe Cactus National Monument reserve, in southern Arizona. It's extremely quiet there because it's a no-fly zone for commercial aircraft and private aircraft; however, the reason

Recording using an ORTF-3D rig, in the Sonoran Deserts of southern Arizona. Military operations in this region can present a challenge to the recordist, so knowing when to visit is important. (Matt Mikkelsen)

it's a no-fly zone is because it's a military operating area, so the airspace above that part of the US from 9 a.m. to 5 p.m. on weekdays is incredibly noisy. There are fighter jets all over the place. For the most part, though, it's so quiet there and you have the beautiful sounds of the desert surrounding you (Audio 10.6), especially overnight and on weekends at dawn. You just have to know when and how to work around everything.

What are your approaches to logistics and planning, especially with longer recording trips?

I've been on quite a few long recording trips (up to three-and-a-half months, in some cases). There's a lot of driving and walking involved, and you need a few rest days here and there, just to recharge. On some trips, I might go to a location confident that I'll only need to spend a weekend there; then I'll show up, realise it's not suitable and move on. The opposite can happen too: sometimes I'll randomly find places that weren't even on my map and end up spending two weeks there. I usually have a rough idea of the ecosystems I'd like to explore or groups of species that I'd like to focus on before I start out, but ultimately, it's a game of patience and perseverance and you have to spend a lot of time in the field to get what you want.

Matt follows a trail through the temperate coniferous rainforests of Olympic National Park. (Nick McMahon)

There are some cases where I need to do a lot of research beforehand. If I'm traveling to a place I'm less familiar with—like the Amazon (Audio 10.7) or Costa Rica (Audio 10.8), for example—I'll do a lot of reading about the species I might encounter and what their behaviours are like. Going with the flow is really important too: with this kind of work, you have to be comfortable not really knowing when/where you're going to sleep each night, or when you're going to have your next shower or hot meal. When it comes to recording wild species and their habitats, you never know when you're going to get that once in a lifetime opportunity, so you always need to be ready and you have to be constantly focused. That can be incredibly draining.

In the midst of all this travel, I'm very aware that going to the likes of the Amazon and Costa Rica is a privilege. These are not inexpensive places to go to: there are plane tickets to buy, expedition costs, and there's a carbon footprint to be aware of. You also need time, and that's one of the things I'm very lucky to have at this stage. If you have lots of care commitments or have to look after kids, it's very difficult to go on recording trips of any kind.

You are known for your focus on remote habitats, but you've also recorded in urban and suburban environments. What wildlife sounds have you encountered in these more populated areas?

I did a very targeted urban soundscape recording project a while back, where I spent a couple of months travelling to 13 different cities around the USA,

Matt, recording waterfalls in the gorges of central New York. (Palmer Morse)

recording the most iconic sounds of each city. It was a really informative trip, because I discovered that there's so much cool stuff happening in urban green spaces. In New York, for example, all the wildlife that calls the island of Manhattan 'home' has less space to move around, so species tend to congregate in the likes of Central Park. You can sit on a bench in Central Park, and in 20 minutes, you might see 20 or even 30 different bird species. That generally doesn't happen in remote places because the animal activity is much more spaced out and open.

Despite that fact that I focus on remote soundscapes away from human settlements, I love hearing what we, as humans, have created: what we sound like and what our built-up environments sound like. We are a part of nature—a species that has evolved on this planet—so thinking about the sounds we make alongside wildlife is interesting and it helps us gain more context on issues like noise pollution.

What were your recording approaches for the urban soundscape project?

I was still using my ORTF-3D rig, which doesn't exactly blend in very well in a city! At some point early on in the project, I bought myself a hi-vis construction vest and hard hat. I would walk around the cities with my recording gear, looking like a construction worker or a surveyor. If I didn't do this, I found that people would come up to me and ask, 'Who are you working for? Is that a camera? What are you doing?' And they would ruin these beautiful recordings of the cityscape. But when I put on a hard hat and a hi-vis vest, people just walked right past me.

So, that's my pro tip for recording in cities! I know other people who put signs that say 'recording in progress, please don't interrupt' or similar on their microphone stands. But I've found that this can influence people's behaviour: if they know they're being recorded, they're not going to talk or they're going to act differently.

What would your main points of advice be for recordists curious about exploring wildlife and remote natural habitats?

When recording wildlife in remote locations, the first thing to remember is personal safety, especially if you're working alone. You should be confident working in isolated environments and understand what the risks are. Don't be so excited to go out and record the dawn chorus that you forget to take essentials like water or food.

I also learned very early on that sound recording is not about having great equipment or travelling to exotic locations. Being a good listener is much more important and it's a skill that you have to practice continuously. A lot of people think, 'Once I get all of this this fancy audio gear, I'm going to plan a trip in another country and I'll get such great material.' But that's not the case. Don't wait to get fancy audio equipment: that will come. Start by going outside and listening.

On a related note, the other big piece of advice I have is that you need to spend lots of time recording, and I've already mentioned how lucky I am in this regard. An analogy with photography might be that in order to get that one incredible

Recording the vast prairie habitat of Eastern Montana. (Matt Mikkelsen)

shot of a wolf taking down an elk, you need to shoot probably 10,000 or 20,000 photos. You need to spend time doing it and practising it. It takes a lot of effort to work on your craft and record those really special moments, and it's never-ending.

Finally, one of the first things Gordon Hempton told me when I was starting out as a recordist was, 'Don't think, just feel.' He said that to me all the time: 'Don't think, just feel, don't think just feel.' I'd be standing there holding $10,000 of equipment and he'd say that, over and over. And I was asking myself, 'What do you mean, "Don't think, just feel?" I'm standing here holding all this stuff and I need to set my gain correctly and the position of the microphones is wrong!' But he was right. The more time you spend paying attention to your feelings, the less time you spend thinking about the recording process and letting equipment get in the way of the natural behaviours you are trying to document. I try to keep that in mind each time I press record.

Laura Giannone

Laura Giannone's multifaceted work encompasses everything from weather and thunderstorm recording to acoustic monitoring and ornithological surveys. In this interview, she discusses topics as varied as plant/vegetation sounds, equipment customisation and weatherproofing, Coyote tracking, hummingbird flight activity, the behavioural tendencies of wind, and the fascinating ecology of the Mojave Desert in the USA, where she is currently based. She closes with some thoughts on the significant challenges facing female and gender-expansive audio professionals. *MF*

How did you become interested in wildlife sound recording? What inspires you and why do you do what you do?

I've always loved being in the natural world. From a young age, I've enjoyed documenting the environmental impacts of humans (my earliest experiences in this area involved monitoring watersheds). I also love troubleshooting in difficult weather conditions, doing weird crafts by hand, modifying equipment, and activities like mapping. The practice of sound recording calls all of these elements together in a way that suits me.

I used to live in a rural area which was relatively free from human-induced noise. Some friends there worked in wildlife conservation, studying birds. I learned about birding from them and gained insight into how listening formed a part of their surveys; at the same time, I became really interested in learning avian vocalisations by ear. I was also frustrated by the increase of anthropogenic noise over the area and wanted to learn how to document noise pollution. All of these elements have informed my development as a recordist and my current work in the field.

You have a personal interest in weather, particularly storm chasing. What is it like to record a storm and what recording methods do you use?

For me, thunderstorm recording (Audio 11.1, 11.2) is mostly about the adrenaline rush. I enjoy studying forecasts and historical weather data, as well as learning about the specifics of a location, like how topography might influence weather or

Laura follows a developing storm from a parked vehicle. (Laura Giannone)

This photo was taken as Laura monitored storm development. Mics (not pictured) have been positioned nearby. (Laura Giannone)

any hazards that are present. Setting up can take hours and most of the work is about trying out various field positions, as well as choosing suitable materials for shielding microphones from wind, rain and hail. There's lots of tweaking of cord and tie-off points, lots of swearing! I do most of my thunderstorm recording with Matt Mikkelsen, who is a fantastic nature recordist as well as a good friend and field buddy. Recording with someone else is safer, more fun and better all-round. I learn lots working with Matt and he inspired me to pursue thunderstorms as a recording subject.

My main workflow involves assessing weather and location, choosing sensible recording positions (i.e. areas that are far from vehicle noise, accessible in heavy rain, acoustically intriguing and with some stands of trees nearby), then setting everything up and taking cover in a vehicle while lightning is active. Learning about weather and safety is key. Gear maintenance is critical, especially on longer trips: rodents chew cables and stuff gets wet, so I have drying-out routines to keep ahead of things and I always bring some spare audio cables with me.

You have been heavily involved in acoustic monitoring and habitat surveys. For example, you have undertaken survey work in the Pacific Northwest, using passive acoustic monitoring (PAM) techniques and technologies to explore the impacts of land management practices on avian species. Can you

describe some of the basic elements of this work and the recording methods used for it?

Acoustic monitoring can help bolster data about species presence, and this can help determine which (and when) species use particular habitats. One study I helped with involved the collection of data about which key bird species were present in old-growth and clear-cut stands of trees. Once enough audio data was collected using recording devices placed in these locations, it was analysed alongside habitat data. Various organisations could then use the information to assess the potential impact of land management strategies on wildlife.

In terms of the processes and approaches behind this kind of PAM work, I'll often attach recording devices to trees according to a survey plan, which might involve following plots at various stages of timber production or covering transects within a survey area. It's important to consider distance from flowing water (which can obscure acoustic transmission of birdsong), as well as how securely things are attached, ease of access, seasonal changes and variations, and other factors. Survey devices will typically run at dawn and dusk to cover peak avian vocalisation periods, and there will usually be an agreed schedule for battery replacement and data offloading.

For some acoustic monitoring work, things like humidity and the remoteness of survey sites will affect design considerations for the devices being used. In one location, I soldered additional battery packs onto devices to help with powering and made a few other adjustments for better weatherproofing. These modifications allowed the devices to stay in place for longer, which really helped with the survey. It's not always easy to customise things: some devices are more suitable for modification than others, and when changes are made, it's often necessary to do some extra weatherproofing. If there is an option and budget for it, more expensive products which have robust weatherproofing options included at the point of manufacture are ideal. Many acoustic monitoring devices on the market at the moment have either built-in weatherproofing or accessories to add, such as plastic boxes which latch securely and have gaskets incorporated for mics. Strong cords and/or sturdy straps are great for attaching recorders to trees, but there is always the possibility that curious wildlife will have a look at things, which is something to bear in mind.

When it comes to data analysis, there are currently a range of processes and programs which can help speed the whole process up. Some software packages have template-matching capabilities, where users create templates for species and then search for recordings which fit. This really helps when processing many hours of acoustic data.

How do you listen to and record species? Do you start with a wider sonic frame before narrowing down and focusing on individual sounds, or the other way around?

Species-focused work is a bit of a luxury for me, so I'll often record individual vocalisations opportunistically. For example, if I notice something exciting or find some clues about animal behaviour during my regular recording activities, I'll position a drop-rig out to make an unattended recording. On other occasions and if the opportunity allows, I'll target specific subjects (Audio 11.3). I'll also observe my surroundings to figure out sensible approaches and avoid getting too close.

With Coyotes, for example, I might pay attention to where their scat accumulates, what they're eating and how these elements change during seasonal and weather shifts. Understanding the life cycle of foods such as Coyote Melon might be relevant in such cases, since certain seeds could be present in their scat. I'll also note when and where the Coyotes call at night and dawn. If I see any individuals, I'll pay close attention to where they're going and where they're coming from, studying tracks and other signs of animal movement. Even with

Pieces of plant matter, fur and bones, present within what is likely Coyote scat. Paying attention to what wildlife is eating can provide useful indicators about animal behaviour, as well as the flora and fauna present in the wider environment. (Laura Giannone)

all of this observation, my favourite recording of howling Coyotes might end up coming from a drop-rig I left overnight, during a thunderstorm recording in a completely different location (Audio 11.4).

It's also interesting to explore non-vocal sounds that species make. Many birds forage on wooden surfaces, tapping and drumming for invertebrates. Grouse species make really interesting sounds with their wings, and some birds toss around vegetation. Branches, faeces and urine cascade down through dense foliage due to primate activity in tree canopies, and sometimes primates will throw things. Elk stamp and make all kinds of interesting sounds as they move. Insect wingbeats are fascinating to listen to. I also enjoy learning about the dietary preferences of living things, which helps with recording in general. Local guides are great resources for this, combined with field observation.

Seasonality and time of day are key considerations, as well as background reading and local knowledge. If I'm interested in birds that utilise palm trees for feeding or displaying, for example, I'll think about going when they'll be most active; on the other hand, if I want palm frond sounds on their own, I'll go there at night when the birds aren't active. I read ornithology books and online resources, and look at migration data. I also ask questions and listen to stories from locals and friends about animals they know well, or things my neighbours see or hear. Sometimes, a neighbour or friend will tell me they saw a certain migratory bird coming through, or maybe a Bobcat. Learning from locals is best, where possible. I also really enjoy working with field guides, learning from them and following their expert guidance on how to be aware and respectful of local ecology.

Another element to be aware of is recordist presence. Sometimes, my activity on a trail will set off birds' alarm calls; if that happens, I'll consider leaving mics out for longer durations, either moving myself away from the area or staying very still for a long time. I bring binoculars with me where possible, so I can spend time observing species from a distance. It's always nice to confirm a species' ID visually and make notes about appearance or behaviour, so I can learn more about it later.

What sounds have you recorded around your local area in the Mojave Desert? Are there any unique fieldcraft considerations with this location?

The high desert ecology is so diverse. I love learning about the topography, microclimates, plant populations and movements of wildlife here. I especially enjoy learning about plants and how they change over the course of the year (including where they might propagate and what species interact with them), as well as how seasonal water flow affects topography, seed dispersal and root development. Indigenous people have used many plants in this region for thousands of years, and learning about traditional plant uses is super helpful for understanding where and why certain plants may grow. Some plant species have been introduced through colonial cattle ranching, which has unfortunately contributed to fires spreading.

Vegetation sounds can vary considerably based on growth/flowering stages, dryness and other factors. (Laura Giannone)

In terms of my recording practice, understanding plants really helps with targeting sonic textures, especially where wind is concerned: for example, the sound of wind moving through and across grasses (Audio 11.5) is very different from how it sounds moving through Ocotillo and sage species (Audio 11.6). Plant structure sounds also change as foliage grows and flowers; when flowering takes place, it attracts pollinators and this leads to different sounds again. Dry grasses sound very different to short or new grass growth. I've visited some locations which have reopened after wildfires, and it's really interesting to observe the changes in recovering plant communities over time there, as well as their corresponding soundscapes.

Wind is fascinating and ever-changing: I could spend every day of my life recording sounds generated by the wind and there would always be something

Two contrasting photos, showing Joshua Trees in the Mojave Desert. The first photo (left) shows a Joshua Tree growing normally; the second photo (right) shows the species following a wildfire. The soundscape of a burn area can vary significantly from untouched portions of the landscape. (Laura Giannone)

new to discover. The sounds that wind causes will vary depending on wind direction, humidity, temperature, the growth stages of plants, and the position/types of rocks and other topographical features throughout the landscape. Even how loose or compacted the sand is can have an effect on wind sounds, and the structure of sand itself is shaped by weather influences. I remember listening to the dried-out stalk of a plant which had a hollowed vascular system, rolling around on the side of a sandy slope in the wind: as it moved, it sounded a little different depending on where it was rolling, how strong each gust was and how easily the sand scattered.

There are lots of hummingbirds in the Mojave and I enjoy being dive-bombed by them. Costa's Hummingbirds are precocious, and it's pretty fun when they get curious about mics and hover around them (Audio 11.7).

What challenges have you encountered in your career and how have you navigated them?

Sometimes, people say or write that there aren't many women in sound. This concept is an attempt to shift blame and further isolate women from each other. The professional audio industry is a massive field with many types of work and skill-sets. It's not that women don't exist within it: we've had to work just like anyone else, but with the added obstacles of being assaulted, discriminated against, harassed and discouraged in so many ways.

Women and gender-expansive people have had to overcome more obstacles and challenges than normal to achieve positions in employment. This is a direct result of discrimination, and we are sometimes abused to the point of having to find other ways to pursue our passions and use our skills. We may have to turn down opportunities because there are unfair strings attached, or we may be denied those same opportunities as a result of bigotry, rather than the myth that 'we don't exist'.

Laura prepares a double mid-side rig for thunderstorm recording. (Laura Giannone)

Self-awareness in recovery from trauma—and community support—are so important. I am very fortunate in that I work with (and for) lovely people, who are supportive of each other and genuinely care about each other's wellbeing. The wonderful people in my work and personal life help me get through bad experiences. More widely, members of the field recording community generally support each other through shared passion for listening. There are so many inspiring women, gender-expansive people and male allies in the wide world of sound.

Seán Ronayne

In 2021, Seán Ronayne set out to record every commonly occurring bird species on the island of Ireland: a task which, at the time of writing in early 2025, is almost complete. After chatting to him and taking time to listen to his work, it is clear that these kinds of ambitious projects are underpinned by a deep love of the natural world and a heartfelt respect for wildlife which runs all the way back to his earliest childhood experiences. We explored his influences and motivations, discussing fieldcraft and technical approaches, mimicry, nocturnal bird recording, and his wider thoughts on nature and wildlife conservation. In the closing portion of the interview, Seán wrote openly about his late diagnosis of autism, ending with some personal thoughts on family and the importance of experiencing the natural world first-hand. *MF*

How did you become interested in wildlife sound recording? What inspires you and why do you do what you do?

I was always a hyperactive kid, and my parents learned very early on that immersion in nature greatly helped with my restlessness. When I was very young, my father would wheel me down to the local nature reserve in my pram and try to imitate the sounds of different birds, mammals and insects, asking me to guess the species. It was a big game to me and I would always laugh at his attempts.

All I ever wanted to do in life was immerse myself in nature and learn as much as I possibly could about it, and this approach has driven all of my work up to the present day. In my school years, I looked out of the window and watched the birds go by. I fantasised about being outside, discovering and learning about wildlife. I wore bottle-green boots and had long hair. Nature class was my favourite and I sometimes corrected the teacher during lessons, which put me in the danger zone! I idolised Éamon de Buitléar, Éanna Ní Lamhna, Don Conroy and David Attenborough. These people did what I wanted to do: they were immersed in nature and they presented its beauty to the everyday person. All of this earned me the rather apt nickname 'Nature Boy', which I was proud of (even though it was used in jest).

After school, I went on to study for a bachelor's degree in Zoology and then did two different master's degrees: one in Marine Biology and another in Ecological Impact Assessment. In 2018, I moved to Catalunya with my partner Alba (a native

Seán takes a break from recording in Mizen Head, County Cork. An acoustic monitoring device was deployed by Seán on Mizen Head for three years, which allowed him to gain an intimate understanding of avian behaviour in the area. (Seán Ronayne)

of Barcelona). I immersed myself in the bird species there and bought my first portable recorder, which gave me a chance to document and further explore the bird sounds I encountered. And that's how my sound recording journey began.

Back in 2021, you embarked on a long-term project to record every bird species on the island of Ireland. How have you approached this project in terms of logistics and recording methods?

Firstly, I needed to decide how many species I would actually record. 'Every species in Ireland' could mean a number of things to people. Approximately 500 avian species have been documented in Ireland, but many of these are one-offs and/or extremely rare migrants. It would be impossible to record all of them. So, I decided to focus on regularly occurring Irish birds: resident species, those that come here for summer/winter, and any using Ireland as a stopping point on their migrations farther afield.

For me, recording methods aren't overly important: I just throw my energy into whatever it is that I decide to do. I began recording [Irish bird species] in my parents' garden when we moved back from Catalunya in late 2021, using a tripod-mounted parabolic reflector pointed towards a bird feeder. Even though I was recording during the COVID-19 lockdown, it was wonderful because I could watch the comings and goings of birds from my childhood bedroom window. I eventually got a COVID travel permit for bird survey work and began adding more and more species to my library, setting up my reflector and some PAM devices around various survey sites. The passive devices ran for six hours each day, and each night I would analyse recordings from all of them. I began saving sounds into ordered folders, creating a list of the species I needed to record and ticking them off, which I got a real buzz out of (Audio 12.1).

I soon invested in two better-quality PAM devices with increased weather-proofing, leaving one at Tacumshin in County Wexford, and another at Lissagriffin on the Mizen Head Peninsula, which is Ireland's most south-westerly point. I set the units to record from dusk to four hours after sunrise, every night for three years. This supplied me with detailed information on the species present in each area, including non-avian species such as Otters (Audio 12.2). It also helped me to understand when particular birds migrated, when their activity peaked, what time of night they tended to fly over at, and so on. I made thousands of recordings this way, counting every single call, assigning it to species and vocalisation type, then dating and timing each occurrence. I stayed up until two or three o'clock each morning, analysing data from my monthly collections: something I had to do, otherwise I would have drowned in sound files! It was so much work that I ended up with severe pain in my knuckles from constant mouse-clicking and typing.

By the third year of the project, when I had approximately 180 of the 200 regularly occurring avian species recorded, most birds left on my list were rare, not very vocal, far-ranging, or difficult to approach (in various combinations).

Seán, recording seabirds along the cliffs of Tory Island in Donegal. He has also recorded Corncrakes, the aerial 'drumming' display of Common Snipe and Eurasian Skylark song in this location (see Audio 12.1). (Alba Novell Capdevila)

Seán uses hydrophones to listen to a rock pool in east County Cork. (Alba Novell Capdevila)

Because of the rarity and protected status of these remaining species, I needed to be licenced and accompanied by National Parks and Wildlife Service rangers on certain sites. With hard work and a lot of perseverance, at the time of this interview I have recorded 199 Irish bird species and have plans to record three more (Grey Partridge, Great Skua, and Red-breasted Merganser) to complete my mission. I hope to finish the project soon.

Have you made any interesting discoveries over the course of the project?

My recording project around Ireland provided many opportunities to showcase the beauty on our doorstep and highlight how it is disappearing.

One bird which came to symbolise this was the Ring Ouzel: a close relative of the familiar Common Blackbird. I knew Ring Ouzels were rare in Ireland, but when I enquired about locating them, I was very upset to learn that the country was down to just a single pair. I was so frustrated about the species' conservation status that I made it an immediate mission to record its sounds (Audio 12.3). These are, sadly, the sounds of extinction on our very own doorstep and I now use them as a potent reminder of what we are losing.

There were also many positive surprises. Because I had autonomous recorders out in the field for years at a time, I was able to pick up sounds from some very rare and unexpected visitors, including a Semipalmated Plover [endemic to North America]. This bird was detected for the first time in County Cork via my autonomous recording devices; there were a handful of Irish records for the species at the time, but the only proof of its appearance in the county was held in my audio files for several months until it was observed and confirmed in the field by other people. Long-duration passive recording also provided copious amounts of data on the migratory patterns of birds, revealing that certain species—once thought extremely rare in Ireland—were, in fact, migrating overhead at night in reasonable numbers. Tree Pipit is a good example: it doesn't breed in Ireland and typically only turns up as a rarity in spring and autumn. To see one every year is good going, but at times, my devices were picking up several each night. This shows how much we miss during surveys and how much we still have to learn.

Vocal mimicry has been another key area of interest for me during the project. I often angle a tripod-mounted parabolic reflector towards the sky at night, to monitor the nocturnal calls of migratory birds. After noticing a Common Starling nesting along the eaves of my house, I decided to alter the position of my nocturnal reflector set-up slightly, so that it gathered sounds from the edge of the roof as well as overhead; in this way, I was able to record the starling's song each morning, in addition to any migratory calls from other species I had documented overnight. Common Starlings are lifelong learners, but many birds are not: they have a critical learning window of a few months, after which their song is firmly established. I wanted to know if I could hear the starling add to its mimetic repertoire through time, and that's exactly what happened. It began to add sounds produced by

Seán stops for a rest with Toby, during a recording trip in Catalunya. (Alba Novell Capdevila)

nocturnally migrating birds to its own song, including passing Common Snipe and Golden Plover. And it wasn't just birds: when foxes began to mate in February, the starling added their shrieking sounds to its vocabulary. Amazingly, it even started to mimic me! It began to sing 'c'mere' [come here] in my Cork accent, which is how I call for my dog Toby in the back garden.

On another occasion when walking Toby through a local woodland after work, I heard a dog barking just ahead of us. As I crouched down to put Toby's leash on, the barking started again: this time from the top of a large oak tree. Baffled, we both looked up, only to see a Eurasian Jay hopping up and down on a branch, eyes fixed on Toby. It was barking at him like an angry Labrador! Assuming nobody would believe that I had encountered this behaviour, I quickly recorded it with my phone, thinking it was new to science. I rushed home to write it up, only to discover that it is a well-known phenomenon called context-dependent vocal mimicry.

On a much wider note, after a lifetime of listening to the natural world, my main discovery has been that I know very little about it. Whilst it probably sounds a bit poetic to say this, it also comes with an element of realism and sadness. Many species will leave this world before humans ever get a chance to understand them. Humanity is currently facing a planetary extinction event many times greater than we have known, driven by our actions as a destructive, profit-driven mammal. I genuinely believe that most of us want a nature-filled world, but we live under systems fuelled by profit and growth which detrimentally affect us, our neighbouring species and our wild spaces. If enough of us fight for change, I believe we can halt and reverse our damaging ways.

Can you give me an overview of your recording approaches? Do you have any examples of unusual or interesting field situations, and how you adapted your approaches to deal with them?

My approach to recording is fundamentally non-technical; I don't have a background in audio, so I put in relentless effort (with a lot of failure) to get what I need. I would describe this as 'success through respectful microphonic bombardment!'

Making recordings without disturbance is crucial for me, so I take every precaution to avoid annoying or distressing birds. I observe target species in detail, moving carefully and keeping my distance as much as possible: then I place my mics and get out of the way. I prefer leaving equipment unattended for long periods of time, because it means I don't get in the way of the birds and their natural behaviours.

In terms of unusual situations, on the Irish island of Skellig Michael (when recording for my album *Wild Silence*), I rigged miniature omnidirectional mics on vegetation near Puffin burrow entrances (Audio 12.4), and lowered mics over the edges of cliffs to get close to Razorbills and Northern Fulmars. I also left several autonomous recording rigs in interesting locations, and used a hydrophone to

A Puffin, photographed on Skellig Michael. Mini omnidirectional mics were mounted near the entrances to Puffin burrows on the island, to record a range of behaviours in detail. (Alba Novell Capdevila)

Seán, recording in stereo with a bisected-type reflector. (Chris Maddaloni)

record some fish sounds to contextualise my other work. Because of the species and habitats I was dealing with on Skellig Michael, everything had to be conducted under licence.

How does nocturnal recording (e.g. of migratory birds) tie-in with your other work? Are there any key differences?

Recording nocturnal bird vocalisations was my gateway drug into sound recording!

I've been listening to birds for as long as I can remember, but I only started recording them after attending a talk by Magnus Robb, in the Ebro Delta of Catalunya in 2019. Magnus' talk literally transformed my view of the audible world, because he explored some basic approaches and techniques for nocturnal sound recording and explained how much we could learn from the processes behind it. I was living in Barcelona at the time, and when I returned to the urban sprawl of the city after the talk, I started using a portable recorder (placed inside a plastic measuring jug covered with cling film and secured with an elastic band) to record birds passing over the roof of my flat as I slept. With this simple solution, I began to pick up all kinds of out-of-context migratory species sounds at night.

It was a complete revelation: all of a sudden, the densely populated human world around me became a place where a multitude of interesting bird sounds could be experienced. I was able to listen to flocks of flamingos flying over high-rise apartments at 3 a.m., far away from their typical salt marsh habitats. I could also hear Ring Ouzels passing by. After some time using the portable recorder and sealed jar approach, I upgraded my nocturnal recording system to a parabolic reflector set-up, which enabled much clearer audio and helped attenuate the various noises of the urban sprawl below my roof (Audio 12.5).

Your book, *Nature Boy*, explores your personal connections to wildlife and your journey as a naturalist/ornithologist. Based on the book's themes, it is clear that listening to wildlife is a deeply personal activity for you, rooted in your childhood experiences of nature. Do you have any thoughts on why you decided to record sounds from the natural world and what wildlife sound recording means to you on a personal level?

Nature means everything to me. It's what makes me feel whole. It's what drives me, what I dream about and what I need in my life every day. But the sounds of nature are particularly special and connect with me on a deeply personal level.

In my thirties, I had a late diagnosis of autism. My wonderful partner Alba was the one who solved this puzzle, having studied autism for her final dissertation. I've always had serious personal issues with anthropogenic noise: traffic, planes, loud music systems, construction activity, the shouts and screams of drunken revellers. These kinds of sounds make my skin crawl and raise my anxiety levels through the roof. Lots of people don't like these sounds, but for me, they have always caused extreme levels of distress, to the point where I can no longer function coherently and I mentally black out with stress or panic. My diagnosis brought so many insights and flashes of understanding: all of a sudden, everything made sense. Even my earliest nature experiences with my dad wheeling me along in the pram made sense, because I could see how the calm and tranquillity of natural spaces helped me. From childhood, it's clear that I was seeking comfort in the sounds of wildlife and the natural world.

As I write this, I'm visiting family in Barcelona, after a very busy year spending too much time in urban areas for work-related reasons. A few days ago, I travelled to the high Pyrenees of Catalunya. The mountains were cloaked in snow, so I had to leave the car in the last accessible village and walk to the top of the nearest mountain: a three-hour hike. I was vastly unprepared, with just a pair of jeans, a few thermal layers and a hooded top. But that didn't matter: the farther I walked up the mountain in the dense snow, the less human activity I heard. I left behind so much stress and anxiety with each step; I felt eight years old again, where I belonged. I could hear the blood rushing through my head, the gentle sound of falling snow, creaking pine trees and a plethora of birds (Audio 12.6). I was in a state of bliss for the whole day, despite icicles forming on my nose and hair! I realised there and then that I belong outdoors, immersed in natural environments, blanketed by their wonderful sights and sounds. After my experience in the Pyrenees, I promised myself I wouldn't allow work to take me away from these places, and I intend to keep that promise.

Seán, with his partner Alba and her dog Dori, in the Catalan Pyrenees. (Seán Ronayne)

I don't want my little daughter Laia to grow up always surrounded by our fast-paced ways of living. I want to nourish her with the true tonic of nature. Yesterday, I took her to a local House Sparrow roost in our family's Barcelona suburb. There must have been 500 birds in the pine trees, calling so loudly that they drowned out the surrounding traffic (Audio 12.7). They owned that airspace in the most beautiful way.

Laia heard them too and gazed up with innocent curiosity. I told her that one day, we'd come back to have another listen. And we will.

Hans van den Berg

In his own words, Hans van den Berg spent 'many happy hours outside as a child, barefoot in the South African dust, climbing trees and exploring [his] surroundings'. His wide-ranging, intuitive understanding of African wildlife and habitats became clear as we discussed a range of topics, including: logistical and fieldcraft considerations around Kruger National Park and the Central Kalahari Game Reserve; recording approaches for Common Barking Geckos, Black-backed Jackals and other species; and the influence of seasonal and meteorological phenomena such as the El Niño–Southern Oscillation (ENSO) on animal behaviour. In the final two responses of the interview, Hans shared his personal thoughts about human impacts on nature and the ever-increasing presence of anthropogenic noise in his adopted UK homeland and beyond, echoing the concerns of other wildlife sound recordists globally. *MF*

How did you become interested in wildlife sound recording? What inspires you and why do you do what you do?

My parents encouraged and nurtured my love of the natural world, and instilled a deep respect for it. I recall spending many happy hours outside as a child,

The male Lion encountered by Hans in Botswana. See Audio 13.1 for the associated smartphone recording. (Hans van den Berg)

barefoot in the South African dust, climbing trees and exploring my surroundings. It's difficult to articulate how special Africa is to me: it's my spiritual sanctuary, and my inner compass points there no matter where I go. I love listening to wildlife in England, but for me, the sounds and experiences of the African bush are truly remarkable and they'll always be a part of my soul.

I remember the exact moment I got hooked on wildlife sound. It was during a holiday to the central Kalahari of Botswana, when a male Lion roared just two meters away from me as I attempted to photograph him from my vehicle. It's impossible to describe the visceral quality and power of that roar: I could literally feel the vibrations reverberating through my chest cavity, through my bones. I instinctively reached for my phone, knowing that I wanted to record the sound so I could relive it and share it with others in the future.

When I returned home and had a listen (Audio 13.1), I was filled with disappointment. It was nothing like the roar I had experienced. Determined to improve, I decided there and then to learn as much as I could about wildlife sound recording.

You have recorded in Kruger National Park quite a bit over the years. Can you tell me about your experiences there and explain some of the factors involved in recording its wildlife?

Kruger National Park covers a total area of just under 20,000 km^2, which is huge. No national park in the UK compares to it in terms of scale and ecological diversity (there are over 500 bird and 100 reptile species alone). As a result of its size, to get any real wildlife sound recording done you need to use a reliable vehicle, and this comes with its own set of challenges. In certain locations, you'll encounter a constant trickle of cars. If you've stopped to observe or make a recording, this is quickly noticed as a potential wildlife-viewing opportunity; drivers will often approach very slowly in such cases, then speed away when they can't see what you're looking at. Occasionally, people will stop to ask what I'm doing and many a good recording has been ruined by human conversation. You really need to pick your moments and be patient.

In addition to being impacted by climate change, the weather in Kruger, and southern Africa more widely, is shaped by ENSO: as a result, the bush experiences long periods of drought (the El Niño phase) interspersed with periods of rainfall-induced abundance (the La Niña phase). Dry periods draw animals towards water sources, and this helps with recording opportunities; however, as the water disappears, so does the wildlife! It's often boom or bust for the recordist, depending on the climatic situation and its influence on rainfall.

Thunderstorms present wonderful recording opportunities in themselves, but the wildlife behaviours associated with them are spectacular (Audio 13.2). For example, insects can sense the drop in barometric pressure as a storm approaches; as they move out in an attempt to stay ahead of the bad weather, insectivorous

The rainy season arrives across Deception Valley, in the Central Kalahari Game Reserve. Inclement weather conditions can lead to some fantastic recording opportunities, especially with regard to documenting wildlife behaviour. (Hans van den Berg)

Clockwise from top left: Curious primates can tamper with unattended recording equipment; all nine displayed wildlife hazards can be encountered at this picnic site; careful foot placement is as important as microphone placement in the African bush; a four-by-four vehicle fender, following damage by a Lion in the Kalahari. (Hans van den Berg)

birds (who can also sense the pressure change) literally 'surf' the impending storm front, feeding on them. After it rains, there are often huge eruptions of termites performing their nuptial flights, and once again, opportunistic animals take advantage of the protein bonanza: frogs, mongooses, baboons and large numbers of birds in the area will come out to feed.

As an example of how significantly weather can influence animal behaviour, on a recent trip to Kruger (during what was predicted to be the end of the El Niño cycle of ENSO), I was hoping to record African Bush Elephants in the Mopane woodlands of the north: a region where elephants are typically abundant. However, due to severe overgrazing of Mopane trees caused by drought, the elephants had moved away from the area and I ended up finding them in significantly higher densities in the grasslands farther south. Many birds had also dispersed to areas outside the park boundaries, and severe thunderstorms each evening ruled out my overnight recording plans.

Time of year can also make a significant difference to the logistics of recording. To use an example from the opposite end of South Africa, the Cape Floristic Region (a UNESCO World Heritage Site in the south-west) typically experiences peak blooms in July and August, with local wildlife adapting to the natural cycle set by the flowers; if you're interested in native avian species like the Cape Sugarbird or Cape Clapper Lark, austral winter is the best the time to go. Migratory birds throughout southern Africa usually arrive by October, but this can vary depending on bird age/species, weather conditions and final migratory destination. In contrast to avian migrants from the Palearctic, intra-African species typically arrive from late October into November. One thing to be aware of is cicadas, which become almost unavoidable in austral summer: they start singing well before the dawn chorus and only stop singing after dark, with minimal pauses. By mid-December, you'll desperately be trying to avoid them, since they can be sonically overbearing and encroach on every single recording (Audio 13.3).

Many species in Kruger pose a significant danger to humans, not to mention vehicles and recording equipment, so it goes without saying that when you visit, you need to be very respectful. In the African bush, you are the guest of other species and you need to act responsibly. Nights can be especially dangerous: after sunset, people should be safely in designated rest camps. Despite the risks, it's a truly spectacular place which encourages deep connections with nature; some of these connections can literally change the course of people's lives and instil a deep love and respect for the African bush. I think we should also remember that in a global context, humans pose the greatest danger to the safety of others.

Bearing in mind what you have mentioned previously about Kruger, how do you typically record there? Do you have any favoured methods or equipment?

The necessary restrictions around Kruger, which I have already discussed, mean that you can only record in reasonably safe areas such as rest camps, designated picnic sites (Audio 13.4), lookout points, game-viewing hides, and vehicles: in other words, in close proximity to people. I employ two recording strategies to overcome these restrictions as best I can.

The first involves the use of drop-rigs, which I typically deploy for overnight recordings around the aforementioned safe spots. If these sites are far enough away from the nearest rest camp, you can get some fantastic human-free sounds from late evening to just after sunrise (Audio 13.5). The second strategy involves recording from within my car, using a pair of omnidirectional microphones in an AB stereo configuration. I'll usually try to position the car in such a way that the vehicle structure acts as a simultaneous windbreak and sound absorber; then I simply hang the microphones out of the window. Sometimes, I place the mics either side of a photography beanbag, which doubles as a head-sized baffle for quasi-binaural recording (Audio 13.6, 13.7).

Recording from a car is challenging for two reasons. Firstly, the vehicle's metal structure expands and contracts as it's driven/parked (and as the ambient temperature changes), resulting in lots of 'pinging' and 'ticking' noises. Secondly, my proximity to the microphone can create unwanted human sounds, which can be mitigated to a certain degree by controlled breathing, restricted movement, and winding the car window up as much as possible.

You have recorded Common Barking Geckos on a number of occasions. What environments/habitats does this reptile favour, and what approaches have you used to convey its sounds effectively?

For me, the distinctive calls of the Common Barking Gecko instantly conjure the atmosphere and sensory experience of the Kalahari Desert: especially dry riverbed zones, where they are often present in significant numbers. The Afrikaans name for the species, *Klipkappertjie*, translates to something like 'little pebble striker', which suitably encapsulates the male's territorial 'kek-kek-kek-kek' vocalisations. I've had the most incredible experiences listening to them after dusk, as the heat from the day disperses from the desert and the moon slowly rises above the Camelthorn trees. One male starts calling, then another and another until the whole area is filled with a chorus of *Klipkappertjies* calling back and forth.

Barking geckos dig fairly deep burrows for protection and males will call from the entrance at night for many hours. Despite this somewhat predictable behaviour, they are a remarkably challenging subject to record well. Finding an

Ideal habitat and recording conditions for Common Barking Gecko, in Botswana. (Hans van den Berg)

area where you have a decent population of geckos can be a bit hit or miss, since they seal their burrows during the day and their performances generally start when most sensible primates should have retreated to a decent-sized tree (or a campsite!) to rest. Overall, finding a good recording location requires a fair bit of time and luck. In my opinion, focusing on a solo gecko close-up doesn't really do the species or its Kalahari habitat justice; recording a whole chorus is subjectively better, and for me it has to be done in at least stereo (Audio 13.8). Approaching things from a soundscape perspective is also the smart thing to do, since many animals in the Kalahari—not just geckos—use burrows to escape the harsh climate. The last thing you want to do is surprise the wrong kind of creature when trying to select a suitable location for a microphone near a burrow entrance!

I'm still trying to make a recording of Common Barking Geckos that I'm truly satisfied with. I keep reminding myself that they are much more attuned to their natural environment than I am, and I always try to bear in mind that my impression of an ideal evening, with all the conditions necessary for a superb performance, will greatly differ from theirs. No two performances are ever the same, which is part of the magic of recording them; at the same time, this creates significant challenges in terms of how variable the resulting stereo image can be, and it's often very hard to determine which configuration (e.g. ORTF, AB) works best.

A few years ago, you recorded a Black-backed Jackal at a watering hole in Botswana. Can you describe that experience for me and explain how you managed to document the species' vocalisations in close-up detail?

The jackal was a surprise recording made at a game farm in the Central Kalahari Game Reserve, where my intention had been to record the elusive, and nowadays quite rare, Brown Hyena.

I had been informed by locals that there was a female hyena in the area who had denned near a watering hole earlier in the season; after a bit of searching, I found the den site and some reasonably fresh tracks nearby. If I left too many signs of my presence at the den, she would avoid it, so I decided to leave a drop-rig at the watering hole and hope for the best. My reasoning with this approach was that I would have a fair chance of recording other species' behaviours as they arrived at night to drink, even if I didn't get the hyena sounds I was interested in.

In terms of general fieldcraft and observation, recording in person near a watering hole used by dangerous animals wasn't an option, for obvious reasons. I also knew that the hyena would stay away if I was nearby: her sense of smell, hearing and nocturnal vision were far superior to mine, so the odds of remaining undetected would be less than zero. Finally, I knew that big cats, wild dogs, hyenas and jackals (most of which would likely be present) have developed a strong instinct to avoid humans at all costs. Taking all of this on board, I decided that using a drop-rig was my only real option.

I wasn't too concerned about leaving my equipment unattended, since I knew from experience in the African bush that it wouldn't attract much attention or cause disturbance; nonetheless, I always wash my hands thoroughly before deploying rigs to avoid leaving natural smells that might entice an animal to have a nibble. Spotted Hyenas are adventurous and will have a go at anything. Elephants can have a mischievous side, while Lions and leopards can be extremely curious. The Honey Badger is a highly intelligent animal with very sharp teeth and strong claws, and they can be remarkably persistent and inquisitive when they find something of interest. Vervet Monkeys and Chacma Baboons are also exceptionally intelligent, with the latter being significantly larger and stronger than vervets: a large male baboon could potentially pose a serious physical threat to the recordist.

With all these elements considered, I placed a tripod-mounted ORTF configuration near the trunk of a small Blackthorn tree, approximately 15 m from the water. The microphones were left out for a few nights. Each morning, I would attempt to 'read' the chaotic animal tracks and imprints left at the watering hole. The news from the ground was always negative: the hyena hadn't visited. So, I would transfer the recordings to my computer and begin analysing the audio files, checking for any interesting behaviours. On one occasion and to make it easier to skim through everything, I turned up the playback volume on my computer; as I strained my ears and listened through my headphones, I heard some low-frequency growling. In my excitement, I neglected to lower the playback level. By the time I realised it wasn't a Brown Hyena, it was too late and I got the full-volume howl of a Black-backed

A trail-camera photo of the Botswanan watering hole discussed by Hans. The pictured jackal (left of image) is almost certainly the one whose calls were recorded. See Audio 13.9 for the associated recording. (Hans van den Berg)

Jackal in my ears, responding to other jackals in the distance (Audio 13.9). It's a very special sound for me, and one of those moments in wildlife sound recording where you are treated to something amazing out of nowhere.

Do you have any favourite recording spots in the UK, where you are currently based? Any species you have recently been focusing on? I'm also curious about how recording wildlife in the UK compares to recording in southern Africa.

For reasons I cannot explain, my favourite recording location in the UK is the north Norfolk coast! I realise that it is as diametrically opposed to my beloved southern Africa as you can get, but there is something about north Norfolk that I love: perhaps the big open skies or the large flocks of geese and waders I remember from watching wildlife shows on Anglia Television as a kid.

On a general note, I find the UK very challenging for wildlife sound recording. The weather conditions are very difficult to work around and, in my view, much less interesting. Like many other countries now, it is virtually impossible to get away from anthropogenic noise: you simply can't avoid roads and vehicles, aircraft noise or other people for more than a few minutes at a time, unless you want to record these sounds alongside wildlife (which some folks enjoy and do very well). Wildlife in the UK also seems to be scarcer than in parts of southern Africa. After hundreds of years of intensive development and cultivation, it feels like most species are now confined to fragmented and ever-shrinking portions of land.

Logistics in Britain are, unfortunately, not much of an issue for me, because nowhere is truly difficult to get to and most areas are relatively accessible. By stark contrast, areas like the Kgalagadi Transfrontier Park (straddling South Africa and Botswana) or the Central Kalahari Game Reserve require lots more planning and preparation. You can't access these locations without a four-wheel drive vehicle, and ideally, you should only go there in a convoy of at least two vehicles. The Central Kalahari Game Reserve alone covers an area of more than 52,000 km^2, which is larger than some countries; in addition, there's no infrastructure other than driving tracks and unfenced campsites with pit latrines, so you need to ensure that you are entirely self-sufficient with regard to food, water and fuel.

For me, these challenges and the remoteness of being in large game reserves are what make the southern African national parks such wonderful places to record in.

Do you have any personal reflections or thoughts on wildlife sound recording that you would like to share?

By nature, I'm quite shy and introverted; I don't like crowded places and I generally steer away from groups of people. That said, I enjoy taking an active role in groups like the WSRS. It has been incredibly rewarding to connect with like-minded people around the world, talk about what we get up to and listen to everyone's recording efforts.

A late afternoon view across Sunday Pan, in the Central Kalahari Game Reserve. (Hans van den Berg)

My introverted nature feeds into my love of the natural world and being in the African bush, although I never really feel lonely there because I regard myself as a part of the ecosystem. In terms of why I record, it keeps me firmly grounded and reminds me of where our [human] place is. I also hope that the sounds I document will help pique people's interest and stimulate conversation about our responsibility for looking after the planet and its habitats.

I don't really want to end on a negative note, but the more I venture outside to record wildlife, the more concerned I become. It is clear to me that we are losing our wild spaces at an alarming and ever-accelerating rate, all for the short-lived gains of a select few. The exploitation of wildlife and Indigenous peoples around the world is abhorrent, not to mention unsustainable; in addition, there seems to be a growing assumption that nature needs to pay for itself in economic terms to be worth protecting. In my view, this reflects not only how poorly we understand our place in the natural world, but our flawed ideology that 'nature' is something other than 'us'.

I only hope this changes and that we begin to treat other species with the respect they deserve. If wildlife sound recording can play a part in that, then I believe it's an activity worth doing.

Andy Martin

Andy Martin has worked in areas such as film, television, radio, VR and interactive audio. As a recordist, he has explored wetlands, prairies, glacial valleys and many other biodiverse habitats throughout his USA homeland and beyond. His interest in wildlife sounds and his work in video game audio are two fundamental and deeply intertwined components of his sonic practice, and they emerged as natural topics of focus for the interview. Around these focal points, we discussed recent trips to the Columbia Plateau and the Atchafalaya Basin, as well as fieldcraft techniques for an award-winning recording of Roosevelt Elk in Olympic National Park. In his closing response, Andy shared some info on future projects, including an upcoming trip to Peruvian Amazonia and the possibility of recording around the Everglades and Okefenokee Swamp in the not-too-distant future. *MF*

How did you become interested in wildlife sound recording? What inspires you and why do you do what you do?

Wildlife sound recording represents the intersection of many paths in my life, all disconnected but generally moving towards each other and sharing themes of exploration and curiosity. As a child, I would spend hours outdoors in natural habitats, observing. I never felt moved to study particular species: just to look, listen, and remember. I remained curious about nature as I got older, reading books about prehistoric and current wildlife, and dreaming of being a zoologist or a veterinarian. Despite these interests, I ended up studying music at college.

Later, I got involved in a video-game sound design project which necessitated recording around my adopted hometown of Seattle. Once outside—listening with intent and not just offhand—I became very interested in the sounds of other species. There's such simple joy in watching birds and other animals come and go, and I realised that when I listened to nature through headphones, I felt the same thrills as when I was exploring outside as a kid. Since that moment of rediscovery, wildlife sound recording has become a huge part of my life.

You have a fascination with the Columbia Plateau's Channelled Scablands region in Washington State, and a deep interest in the species/habitats there. Can you tell me about your recording activity in the Scablands and explain some of the fieldcraft challenges you have encountered?

There is a remarkably complex mixture of recording opportunities and challenges in the Scablands, resulting from interactions between naturally occurring forces and human alterations to the landscape and local environment.

In terms of raw natural forces, the soundscape of the area has been shaped by some of the most destructive and rapid water outflows ever to occur on our planet: huge, prehistoric glacial melts which tore the basalt plain, ripping up boulders, gouging out coulees [ravines] and leaving ravaged canyons and butte-and-basin terrain in their wake. The acoustic qualities of this landscape are wonderful: since there's very little tree coverage, sounds bounce back and forth, producing a wonderful array of echoes. Birds such as the Canyon Wren and Rock Wren take advantage of these properties to harmonise with themselves or stand out in a busy dawn chorus. Common Nighthawks are able to amplify their territorial booms during evening displays (Audio 14.1). Yellow-headed and Red-winged Blackbirds create a raucous cacophony in newly established marshes, which expand in from local waterways. These are fantastic sounds to witness and record.

On the human-influenced end of things, the area's soundscape has been shaped by industrial and agricultural development in recent years. What is now known as the Channelled Scablands region of the Columbia Plateau lies in the heart of Washington State's agricultural zone. Widespread irrigation and power-generating dams have forever changed things there, with sustained inundation (piped in from numerous rivers across the state) used to grow crops such as hops,

Common Nighthawks soar over this spot in the Scablands, their displays and calls echoing off the natural walls of the coulee. See Audio 14.1 for a nighthawk recording made by Andy in this area. (Andy Martin)

apples, cherries, and carrots. As a result of these industrial farming practices, as well as the expanding manufacturing bases and networks of transportation highways which have sprung up to support the growing population in the area, anthropogenic sounds are almost impossible to avoid. Being a dry, relatively treeless and mostly flat region, sounds travel far and I've found that certain locations in the Scablands can be recording wonders one day and sonic nightmares the next. Each visit necessitates long hikes and listening sessions to decide if/where to record.

Human influences have also led to the introduction of invasive species such as the American Bullfrog (Audio 14.2). Prior to 1850, American Bullfrogs were virtually unknown west of the Mississippi River, but their large size and hardiness has contributed to their propagation throughout the west in states such as Washington, Oregon, and California, where they are outcompeting native frogs and even proving to be a negative predatory influence on birds and small fish.

Despite these drawbacks, some superb recording opportunities have emerged as a result of human activity. Sandhill Cranes stop by in their thousands for a few weeks every spring and autumn, as part of their annual migrations between their summer breeding grounds in Canada and Alaska and their wintering grounds in California. New agricultural fields provide vital resting and feeding grounds in the middle of their journey to enable them to rebuild their energy stores and

Andy's favourite coulee, in the Columbia Plateau's Channelled Scablands region. Visible on the cliffside are multiple layers of basalt, laid down over thousands of years and exposed by prehistoric floods from glacial Lake Missoula. See Audio 14.3 for a summer dawn chorus recording made around the pictured coulee. (Andy Martin)

rest before continuing on. With many rivers now diverted for irrigation, perennial streams are often blocked and water is stored in vast reservoirs. This reservoir water seeps through porous basalt, only to rise back to the surface through numerous potholes and coulees farther south. The resulting 'seep lakes' form part of the Columbia National Wildlife Refuge and adjoining Seep Lakes Wildlife Area: very attractive spots for millions of migratory birds and, of course, the wildlife sound recordist (Audio 14.3).

This mix of challenges is exciting for the wandering listener. I find myself endlessly fascinated by the unusual sonic combinations of invasive American Bullfrogs, relocated blackbirds, native Canyon Wrens, artificially generated wetlands and bizarre acoustics throughout the Scablands.

The Atchafalaya Basin in Louisiana (the largest freshwater wetland/swamp in the USA) was an area of focus for you in 2022. What were your recording approaches there? Did you have any target species in mind?

In 2022, I was contracted by Montreal-based Compulsion Games to record swamp ambiences from the American Deep South, for the action-adventure game *South of*

Midnight. After discussion with the team's audio director, we decided to focus our attention on the Atchafalaya Basin. Home to over 250 bird species and around 65 reptile and amphibian species (as well as a year-round chorus of singing insects), it seemed like the best area to gather a variety of soundscapes and wildlife sounds that would work with the 'Gothic Deep South' aesthetic of the title.

Since seasonal characteristics of swamp soundscapes had to be represented in-game, my recording activity spanned winter, spring and summer. For a project of this duration and size, plenty of planning was necessary. Having never spent time in Louisiana, I had to quickly learn about the history of swamp ecosystems there. I soon realised that while there is a great diversity of wildlife sounds, they aren't evenly distributed across the region, or, as logic would suggest, throughout the year! As a result, I spent many days researching annual species concentrations. After lots of reading and charting, I identified the period between mid-February

A quadraphonic configuration, positioned alongside Little Tensas Bayou in Plaquemine, Louisiana. A few hours after this deployment, an American Alligator began booming from the water. See Audio 14.6 for the associated recording. (Andy Martin)

and mid-March as the best time for recording some of the more typical faunal sounds of the USA's southern states, such as Spring Peepers (Audio 14.4), Carolina Chickadees and various woodpeckers. Mid-April into early June would present opportunities to focus on singing and breeding populations of birds and spring frogs, and late July to mid-August would hold the peak summer concentrations of insects, such as annual cicadas (Audio 14.5).

I divided my recording time in the region into several trips. For the first trip, which lasted four weeks, I spent most of the first week just driving around, ascertaining where I would focus my efforts. I mounted small recorders (equipped with stereo pairs of mini omnidirectional mics) on trees throughout wildlife refuge trails, and on bayou islands; these served as my 'scouting rigs', enabling me to determine what was happening across a fairly wide area. I hadn't thought to hire a kayak or canoe for any amount of time, but a local friend kindly loaned me a pirogue [long, narrow canoe] which I used to get myself farther into the swamps. For my second and third trips, I was able to arrange a boat ride with a local guide, who then dropped me off inside the swamps in areas inaccessible by car or foot. Camping out on a tiny island in the middle of an alligator-filled habitat for several days at a time will stay with me as one of my favourite recording memories!

Andy's camp during his time in Louisiana, on an isolated swamp-island in the Atchafalaya Basin. The pictured area of land was, to all intents and purposes, the only dry spot on the island. Each morning, Andy's guide would check on him to make sure everything was OK. (Andy Martin)

For soundscape recording, I used several spaced quadraphonic arrays (left undisturbed and unmonitored for a day or two at a time), and a double mid-side rig which I actively monitored through headphones. Since I had to be mobile and was often uncertain of the legality of camping in the locations I stayed in, I spent most of my time sleeping in—and living out of—my rental car. Nights proved extremely active, with frogs and insects being particularly vocal. During nocturnal recording sessions as the air cooled and relative humidity spiked, it was necessary to sleep with the windows open, which drenched my equipment with condensing fog! I usually listened to my double mid-side rig via a long cable trailed through the car window, but I sometimes left my vehicle and explored paths late at night with a parabolic reflector, seeking out the sounds of individual insects and frogs. When doing this, I learnt that frogs are remarkably sensitive to vibrations and will quickly go quiet if they feel you walking nearby; if you remain still for some time, they'll start singing again.

Some of my favourite recordings from these trips feature American Alligators. I didn't set out to record their bellows specifically, but I did make sure to set up recorders in areas where I saw indications of their presence. By watching the movement of their red eyes shining back from my flashlight at night, I was able to get an idea of where they were congregating and set my recording rigs in those locations. It wasn't a perfect system, but I got some fascinating material that way (Audio 14.6).

Up to his knees in swamp water and mud in the Atchafalaya Basin, Andy pauses for a selfie with his parabolic reflector. (Andy Martin)

I'm curious about how your involvement in interactive audio shapes your approach in the field. Can you tell me more about that, as well as your general workflow for game audio projects?

Although I've spent the bulk of my career as a video-game sound designer, for my work on *South of Midnight*, I was essentially a sound collector and librarian: the developer's eyes and ears in the swamps, so to speak. A key requirement during my work around the Atchafalaya Basin was to create a library of ambient recordings and soundscapes that spanned winter, spring and summer seasons, and I built my approaches around this. For example, I knew that being present in the mid-winter months would enable me to acquire a relatively silent 'base-layer' swamp ambience which other sounds could be layered over, and I used the appropriate stereo, quadraphonic and double mid-side recording methods to get what I needed. I also knew from my experience in the games industry that by actively collecting individual species sounds (e.g. using parabolic reflectors), Compulsion's audio team would have the freedom to shape and sculpt the audible environments featured in the title by adding individual sounds over the ambiences I had recorded.

I've crafted ambiences for multiple games over the course of my career. In most cases, these are generatively built: individual voices (species calls, songs, etc.) play during the game according to preset rules, with a view to mimicking real-world behaviours. It's generally not a great idea to stream full-length recordings in-game due to console and PC memory constraints, so I'll have a set of mastered recordings which act as templates; from these, I'll edit small snippets which will loop and trigger through the game's audio engine. When mastering my sounds, I focus on everything from fidelity to time of day, season and location, preferring to keep large blocks of similar recordings together. It's very important to make recording notes, whether verbal or written, in the field; these can be really helpful when you reach the cataloguing and mastering stages of a game project.

Other than removing undesirable noises or anthropogenic sounds, very little is done with the original sounds. For individual species material, I deliver two versions: the original with very little modification, and an edit which is completely isolated from its environmental context. The latter type, though unrealistic in bioacoustics terms, is particularly useful in sound design, since it allows developers to utilise the same artificial reverb on the recording as that being used on the rest of the mix. (If a sound effect has a significant amount of its own reverb/acoustics preserved, it may be unusable since it will feel out of place in relation to other sonic material in the game.)

In late 2023, you made an award-winning recording of Roosevelt Elk in Olympic National Park. Can you explain your fieldcraft approach for the species and detail some of the behaviours you were trying to convey?

I love recording elk. They are loud, unpredictable, potentially dangerous and very beautiful animals.

Males gather small harems of three to five females, urging them forward with their characteristic bugles. Being large and heavy, they readily wear down vegetation and their daily migrations leave obvious traces over time. They don't always follow the same track, but they do keep to a generally regular route. From camping and backpacking in the same regions over time, I've learned their patterns and paths reasonably well and have discovered that it's necessary to leave multiple recorders throughout their territory to get interesting sounds. There's no guarantee they'll follow a specific path, though: I've made a number of recordings of small herds quietly walking past my mics! Even so, their telltale footfalls, scraping of antlers on ferns, soft mews (from females) and breathing sounds are fascinating to listen to.

Over the course of the year, males grow massive antler racks covered with sharp points. These signify fitness and strength and are used in competition. You can hear their antlers interlock with loud clashing sounds, similar to tree branches smashing against each other. Unlike Rocky Mountain Elk, the Roosevelt Elk of the Olympic Mountains are shyer and may not engage in conflict when humans are present; that said, they're also more likely to be aggressive to humans and it's a good idea not to be standing nearby when animals weighing 450 kg are battling each other. When following elk trails in search of a local herd to record, I'll occasionally stumble upon a small clearing where ferns have been thrown around and great gashes have appeared in the soil. Branches may be knocked down and rotting logs smashed. These clues are important and help guide my placement of microphones on nearby trees.

For the recording in question (Audio 14.7), I had seen signs that a local herd of elk had been actively moving around a meadow near a ranger's station, and this was later confirmed verbally by the ranger. I set up four different rigs: two quadraphonic tree-ears arrays, with the mics mounted at head height around tree trunks on each side of the meadow; and two stereo sets rigged farther down the trail, at a point leading into the meadow. Quadraphonic arrays are beneficial not just for creating a surround sound recording, but also for ensuring that different angles of approach by species can be converted into stereo. With a stereo set, I might be stuck with all of the action happening to one side, but with four mics, I can simply rotate things in my DAW and assign the side channels to play through front or rear speakers without worry. When using the tree-ears mounting technique, it's good to choose a trunk that's roughly head-sized as a minimum, especially if a quasi-binaural effect is desired; however, because my end goal is primarily loudspeaker playback, I tend to use slightly wider trunks.

After pressing record on each rig and slating, I left everything running with the intention of coming back in two days to retrieve the equipment. Car trouble delayed my return, so I ended up with four nights of material rather than two. When I finally made it back for the retrievals, there were telltale signs of gouged earth and smashed ferns. It looked promising. I excitedly collected my gear and hurried back to my car, then drove to my campsite to transfer the audio files to my laptop. Each recorder had picked up something interesting, but I could tell from analysing the spectrograms that only one rig in the meadow had recorded anything worth listening to.

As I skipped to a section of interest in my DAW, I heard a beautiful deep-roaring challenge. A ferocious battle ensued between two males, lasting for around five minutes. The surrounding trees added a natural reverb, making everything seem larger than life through my headphones. It was amazing to listen to.

You've also recorded Golden-mantled Howlers in Corcovado National Park, Costa Rica. How did you do that and how was it different from recording in more familiar spots back in the USA?

I had some recording companions with me in Costa Rica. We spent a month travelling to four destinations: one in the cloud forests of Monteverde and the rest in Corcovado. We were only able to stay in each destination for four or five nights, so I had to maximise my recording time.

For the first day and night in each place, I spent more time hiking the trails than I did anything else: this gave me a good lay of the land and helped with identifying locations which might prove interesting. For example, I learnt that slow-moving pools of water would be attractive to morning birds and some frog species, while

Typical Golden-mantled Howler habitat in Costa Rica. See Audio 14.8 for the associated recording. (Andy Martin)

faster streams would attract glass frogs and the snakes which pursued them. Insects would follow an expected pattern of thinner choruses during the day, and more orchestrated waves of sound at night. What mattered for me during these initial scouting walks was to learn where sounds would occur and how to recognise signs of wildlife in a land where, frankly, I was a complete outsider.

Golden-mantled Howlers are everywhere in Corcovado, so making a recording of them isn't a challenge (Audio 14.8). Making a recording where you can clearly

Rain in Costa Rica presents a number of challenges in terms of mic protection. To help solve these, Andy first created a belt system to mount his quadraphonic configuration as 'quad tree-ears'. A secondary belt was then used to secure open-cell foam umbrellas above the mics. The umbrellas were thoughtfully constructed: a top layer of foam was used to quietly break-up falling droplets, and a more compacted underlayer acted as a shunt to channel water away. See Audio 14.9 for a recording made using the pictured rig. (Andy Martin)

hear them using vocal fry and lip-popping between howls is much more difficult. Like elk, they don't have a single place they call home: the whole tree canopy is their home. They make their way through it day by day, wandering far and eventually returning, but not in any kind of predictable cycle that allows for easy recording during a brief visit. Equipment choice for this species was my favoured spaced quadraphonic array, placed to the side of trees the monkeys frequented (I didn't want to place my gear directly beneath them in case they dropped waste or shook branches down). I also set up my double mid-side rig a little farther out, to record behaviours from a distant perspective.

What are your plans for future wildlife sound recording projects?

I'd love to travel the world and hear the songs of birds, frogs, and insects from far-flung places.

My recent trips have opened my ears to lots of new sounds: in Louisiana, I discovered the richness of swamp environments and the busy ecological soundscapes of warmer areas; in Costa Rica, I had my first taste of non-northern latitudes and non-Nearctic soundscapes. I also got to experience where North American songbirds go for winter and see primates in the trees. My ears are bigger than my budget, though, and these trips will take time to undertake. Pursuing grant opportunities or finding research teams to coordinate with could make them a reality.

On that note, I've recently partnered with a biology research station in Peru to document bird and insect species on the edge of the Amazon jungle. Once I finish my work there, I'd like to make some return trips to Louisiana, if time allows. I've

A double mid-side rig (just visible centre-left), placed amongst heather to record Coyotes. A long cable was trailed back to a tent hidden in the nearby trees, so all sounds could be monitored. (Andy Martin)

also considered recording around the Everglades and Okefenokee Swamp, to both document these fascinating wetland habitats for myself, and to create sound libraries for use by media consumers and researchers. For now, though, I'm quite content to explore my home territory in Cascadia. For the first time in a long time, I'll be around for the annual Sandhill Crane migration. I'd also like to refine my collection of elk recordings, and I look forward to seeking out wolves, which have only just started to return to our area in the last decade. All being well, there are lots of great experiences ahead.

Helen Hedworth

In this interview, Helen Hedworth provides a glimpse into how wildlife sound recording techniques and technologies are utilised in ecological impact assessments, species population surveys and other scientific projects. Drawing on her experience as a biologist and acoustician, she discusses topics such as: her work with Meerkats in the Kalahari Desert; the impacts of wind farm construction on the Harbour Porpoise; acoustic monitoring techniques; her work on Skokholm Island (UK) with Storm Petrel, Manx Shearwater and other bird species; and perception of sounds beyond the range of human hearing. She ends with a few thoughts on wildlife sound recording's role within scientific research, and how technological developments in audio might help with data gathering and project efficiency in the future. *MF*

How did you become interested in wildlife sound recording? What inspires you and why do you do what you do?

I think my interest in wildlife sound recording came from a love of language, music and natural history. My dad loved languages and there was a wide variety of music played in the family home. He also taught me to identify garden birds and we watched wildlife documentaries together. As a family, we would spend a lot of time at the local zoo, so I became familiar with species from distant lands and the sounds they made. Animals, both wild and domesticated, were an important feature of my childhood.

By the time I was a teenager, I was using dad's old Dictaphone to record the sounds of other animals. The first set of recordings that I remember making were from when I volunteered at The Monkey Sanctuary in Cornwall (UK), in the summer before university. Fast-forward a few decades to the present day, and I've used sound recording for research and surveying, as well as assessments of anthropogenic noise impacts on animal populations. I love that sound can have so many useful applications, but I've also become more aware of noise pollution and its impacts on people and the natural environment; I find it rewarding to be working towards reducing these impacts.

Alongside the commercial and scientific applications of sound recording, I enjoy being out in wild spaces, listening to the soundscape and documenting some of what I hear. I discovered that there was a group of people who also liked to do this, so I joined the WSRS. I've now been involved in the society for more than a decade, as a member or an officer.

You studied Art at university but switched to Zoology (within which you developed an interest in behavioural ecology and acoustic communication). What prompted this change? Was it connected with your curiosity about wildlife sounds?

I did a lot of drawing as a child and always imagined I would go to art school. After my school and college exams, I did a foundation year at university, with the intention of going on to study for an Art degree. But once I started my foundation studies and was able to make art all day long, I didn't know what to do with myself! Seeking inspiration, I went to the library and read books on animal behaviour and biomechanics. This reading influenced my creative projects: I made a documentary film on cat behaviour (featuring my family pets) and another as part of a tented installation on the science of sleep, for which I interviewed people about their dreams. Even though I enjoyed developing my filmmaking skills, I soon wondered if I should change direction.

I discovered that I could take a preliminary year in the biosciences at Cardiff University, prior to a degree in Zoology. It felt like a bit of a wild move, having always considered myself an 'artist' rather than a 'scientist', but the course outline was really exciting. I completed my foundation course in Art and, after spending a month at The Monkey Sanctuary in Cornwall (mentioned previously), I knew that my main interest was in bioacoustics. I just had to study a lot of biology, chemistry and maths before I got to that part! It was a big challenge.

Alongside my studies, I was out checking woodland nestboxes, bird ringing [also known as 'bird banding'] and learning birdsongs. I became the chair of WildSoc (Cardiff University's Natural History Society) and I joined the committee of the local Wildlife Trust group.

You went to the Kalahari Desert as an undergraduate and recorded Meerkats there. Can you tell me a bit about that project and explain some of your approaches for recording their sounds?

This was an amazing year, on the University of Cambridge's Kalahari Meerkat Project in South Africa's Northern Cape. The topic of research was the evolution of cooperative breeding.

Meerkats are obligately social: they must live in groups. Family group members look out for each other and protect and nurture the young (generally the offspring of the dominant male and female). They show cooperative behaviours, such as babysitting and pup-feeding, and perform sentinel duties to protect the group from predation. Understanding how and why they make such sacrifices for the group—giving up valuable foraging time to raise another individual's young or to stand watch while everyone else is feeding—was the overarching theme of the research. Since the project had already been running for over a decade when I arrived, the behavioural interactions of multiple generations of Meerkats had been documented, allowing a detailed understanding of their social and other behaviours, life histories and how individuals were related. They had been habituated to the presence of the researchers, so they weren't disturbed by us and still showed a range of natural behaviours (Audio 15.1).

One element of the research explored their calls. They make a range of sounds, including a contact call, begging calls (as pups), submission calls and alarm calls. When alarming, Meerkats use referential signalling (they use a different call for each predator type) whilst also communicating level of urgency. The Martial Eagle appears to be their most formidable predator, and Meerkats' long-distance eyesight is so good that they would react to seeing one before there was even a black dot in my binoculars! This alarm call led them to go to ground, but sensibly, they wouldn't do this for a Cape Cobra. Interestingly, their typical response to a cobra would be to mob it, lunging and spitting at the snake as a group which usually frightened it away. Sometimes Meerkats were bitten, but they had a remarkable tolerance to the Cape Cobra's neurotoxins and often recovered. Still, not a creature you'd want to share a burrow with...

Once I had learned about different data collection techniques, I was able to take on an individual role and became the sound manager. I delighted in looking after the recording equipment, checking our sound files and managing the sound

Whilst weighing Meerkats as part of the regular routine on the Kalahari Meerkat Project, Helen is used as a convenient lookout post. (Helen Hedworth)

library. People in the field would radio-in if they had a problem with the audio kit or question about the recording approach, and I would give them guidance. After spending every day in the field, it was a luxury to get one day in the sound lab and a happy coincidence that the room was one of just two that were air-conditioned, to keep the computers cool in the 40 °C heat.

We did focal animal studies of pups, where we would put a boom mic near a pup and speak into another mic held close to us. We would record their calls on one channel and describe what was happening on another, using standardised terminology. I also had a chance to work on a project investigating whether the loudness of pup calls influenced provisioning.

Helen collects measurements of Meerkat pup begging calls with a sound-level meter, for research into food provisioning. (Helen Hedworth)

More recently, you've been investigating the ecological impact of wind farm construction on marine mammals in the German North Sea, with a particular focus on Harbour Porpoises. Can you tell me more about this work and how you have used wildlife sound recording for it?

This was work on commercial wind farm projects, and we were carrying out monitoring of sound levels during pile-driving activity. In Germany, wind farm developers are not allowed to exceed a certain loudness threshold of underwater sound (160 dB re 1 $\mu Pa^2 s$ sound exposure level or 190 dB re 1 μPa zero-to-peak sound pressure level at 750 m from source), so we measured it and reported on it within a 24-hour time window. Alongside this, we were monitoring the activity of Harbour Porpoises, the most commonly found marine mammal species in the North Sea and one that has very sensitive hearing. Like all cetaceans in European waters, they are a protected species, making it an offence to kill, injure or disturb them. The Harbour Porpoise also has sites designated to promote its conservation.

We were using PAM to monitor harbour porpoise activity. We programmed autonomous recorders that had hydrophones fitted to them and deployed them alongside noise recorders (to monitor the wind-farm development activity), at specified distances from the pile-driving activity and in a special area of conservation. This part of the North Sea was only around 40 m deep. The noise recorder would be deployed within a frame structure and the porpoise recorder would be

Helen retrieves acoustic data from an autonomous recorder, as part of a PAM study of Harbour Porpoises in the North Sea. (Helen Hedworth)

leashed to the frame, floating above it in the water column. It took a whole team to manage all of the instruments, deployments and retrievals, but despite how busy it was, I loved working with all of the equipment.

The Harbour Porpoise tends to avoid boats and doesn't surface for long, so it isn't easy to spot visually. This makes PAM especially useful for monitoring them, although if the animal is present yet makes no sound, it won't be detected. Harbour Porpoises, however, have been shown to forage for around 60% of their time, using a narrowband high frequency (NBHF) click to echolocate their prey; this means that if they are in the area, they typically won't be silent for long. Interestingly, unlike dolphins (most species of which use broadband clicks for echolocation, and whistles and burst-pulse sounds for communication), the NBHF click is the only sound the Harbour Porpoise makes, for foraging, navigation and social communication (Audio 15.2). The main issue for detecting the species, therefore, is to do with how close they are to the recording equipment: they need to venture within a roughly 300 m radius for hydrophones to detect their clicks, since they are produced at very high frequencies (around 120 to 140 kHz). Studies have also shown that their clicks are not especially loud.

During monitoring, we would take the recordings and run them through software, within which we had built a classifier. This was programmed to identify points of interest in the recordings, by being given the characteristics of a typical signal; it would then extract everything that it thought was a Harbour Porpoise. We would verify these files to throw out the false positives and assess for false negatives. Harbour Porpoises emit clicks in a series called a 'click train', which has a regular inter-click interval but speeds up into a 'feeding buzz' as an individual gets close to capturing a prey item (Audio 15.3). We analysed click trains to determine porpoise activity in and around the pile-driving activity, to work out how long it takes for porpoises to return after the noise has gone away.

You have spent a lot of time on Skokholm Island in the UK, where you have surveyed European Storm Petrel populations. What kinds of considerations do you need to make when using recording equipment to survey this species, and what challenges have you faced when attempting to record the birds up-close on Skokholm?

One of the incredible things about Skokholm is how close you are able to get to wildlife. This removes a lot of challenges for the recordist, but dealing with microphone wind noise can still be very difficult. Whilst the island's introduced European Rabbits and House Mice can be a menace for audio cables, these animals have done some of the summertime seabird visitors a favour by creating burrows. These natural shelters can provide the perfect recording environment: if you are able to get a mic close to the entrance and out of the wind without disturbing the inhabitant, you don't need a lot of gain and can get a clean recording. I find listening to birds in their burrows utterly addictive.

An estimated 2,500 pairs of European Storm Petrel breed on the island, nesting in burrows, rock crevices and drystone walls. Males produce a purring or churring song (often likened to a fairy being sick!), and both males and females make a 'terrr-chick' or hiccupping sound (Audio 15.4, 15.5). There have been a few studies investigating the structural differences of songs and calls between individuals, and a team visiting Skokholm each year makes recordings of the birds to try to gain some insight into this. Given the birds' fondness for nesting in Skokholm's herringbone drystone walls, a new wall was hand-built in a prime location and the monitoring team fitted microphones and cameras into the door of each nest box on the wall to monitor activity. There have been some exciting insights into the private lives of nesting 'stormies' (as we affectionately call them) through this approach. The birds are surveyed multiple times each year using playback, which involves playing a recorded call in order to elicit a response and establish whether

Helen holds a Manx Shearwater—which is about to get a British Trust for Ornithology (BTO) ring put onto its leg—on Skokholm Island. (Helen Hedworth)

A Puffin explores Helen's walking boot with its bill on Skokholm. A handheld Audio recorder is visible in the foreground, with a faux-fur windshield applied to its built-in mics. (Helen Hedworth)

a 'stormy' is in the nesting hole. Not all individuals will respond, however, so a correction factor needs to be calculated to estimate the number of nesting holes occupied.

In addition to European Storm Petrels, nearly 50,000 pairs of Manx Shearwater return from South America to breed on the island each year. They create some incredible sounds: rhythmical, haunting and harsh in flight. In the burrow, it's like they are having conversations with each other (Audio 15.6) and it's a privilege to listen to them. Other inhabitants of interest include: Grey Seals, which haul-out and hang about at the landing area (curious about human arrivals and departures), and whose songs can be heard from the coastal paths; the island's Eurasian Wrens, known to have a different dialect to birds on the UK mainland and other islands; chatty Barn Swallows, which nest in the island's buildings and sometimes end up in a bedroom if a door is left open; and the many gulls that contribute to the daytime soundscape, including the largest species, the Great Black-backed Gull, with its formidable, deep voice.

Has studying other species' vocalisations (e.g. the Harbour Porpoise) and their reactions to sonic stimuli made you re-evaluate your own listening in any way, or rethink how you approach and record wild animals?

Once you start working on or recording species whose vocalisations you're unable to hear with your own ears, it really makes you think. There is a human sonic experience of the world (although there is much research today on how that differs between individuals), but other species interpret things quite differently. Our ears are adapted to capture frequencies that matter most to our survival. We generally lose the highest frequencies as we age and we might lose hearing due to physical trauma but, even without this, we are only able to hear sounds between about 20 Hz and 20 kHz: a very small proportion of life's sounds.

When considering other species, if they are able to hear things we cannot, then they can certainly be disturbed by sounds that don't affect us: even worse, they could suffer hearing damage or injury if the sound is very loud and within the range at which they are sensitive. I often wonder: how long have we been going about our business while creating adverse impacts on wildlife? Thankfully, the issue of anthropogenic noise is very much on the agenda these days, especially marine noise. This is important, because there are lots of noisy human activities at sea: construction of offshore wind farms, ports and harbours (with pile-driving being especially loud); geophysical surveys; and detonations, including clearance of unexploded ordnance dropped during wartime. There is also growing, continuous noise from global shipping channels, which is changing the soundscape of our seas and oceans. This leads to masking of important auditory signals, which could have long-term effects on animal populations by making it harder for species to find food, escape predation or find mates. It's therefore important that we understand how noise is affecting our wildlife and habitats, and reduce the levels of noise that we produce through good design practices and appropriate mitigation measures.

With human-inaudible sounds in mind, one thing I tend to do when I'm recording for my own enjoyment is to use a high sampling rate, just in case there is something interesting going on in the ultrasonic range. This uses up more storage space on recorders, but I enjoy looking for sounds on a spectrogram that I may not have heard: it's like being a sound detective.

Is there anything you would like to see happen in the next 10 or 20 years within your own line of work, with regard to wildlife sound recording? Any thoughts on where wildlife sound recording might go as part of scientific practice?

I think we will see an increased appreciation of how important sound is for many species of animal, and even other organisms such as plants and fungi, which have also been shown to be sensitive to sound or vibrations. We then need to ensure that our activities aren't having a detrimental effect on nature. Noise pollution also has a big impact on our own health: for example, people who live close to noisy roads are at a higher risk of heart disease. Conversely, recent research into how nature improves our health and wellbeing reveals that its sounds may be the most positively potent of all sensory inputs, with complex birdsong right at the top.

In terms of recording, changes in technologies are really exciting. We already have incredibly sensitive microphones and hydrophones, and recorders are able to digitally sample across the full dynamic range via 32-bit float recording. Memory cards are constantly getting cheaper and better, so it's possible to make lots of recordings in a short space of time, at reasonable cost. Then there's the question of what to do with all the data from the recordings we make: here, I find the development of artificial intelligence and machine learning approaches fascinating. If, for example, we can train models to accurately and efficiently identify the wildlife signals we are interested in, we can quickly investigate the geographical distribution of endangered species. Other work on acoustic indices uses the diversity and complexity of soundscapes as a proxy for biodiversity, so sound recordings might also be able to alert us to changes in biodiversity, whether positive or negative.

We might even find that sound recording becomes an important tool for a wider number of scientific projects in the future, supporting conservation efforts globally. And that is a very exciting prospect.

Helen returns to shore, after a humid day working in full PPE for a coastal monitoring project in Brazil. The acoustic recorders used for the project have been safely stored in their carry cases (bottom-right of image). (Helen Hedworth)

Glossary

A

AB – A stereo recording configuration in which two microphones (typically omnidirectional) are spaced apart. The configuration renders spatial information predominantly via time differences (i.e. delays in sound waves reaching the microphone capsules).

acousmatic music – A subgenre of electroacoustic music in which pieces are played back exclusively through loudspeakers, omitting any form of live instrumental performance and often focusing on the raw timbral characteristics of recorded source materials.

acoustic ecology – An area of study concerned with the sonic relationship between humans and the environment. Also known as soundscape studies or ecoacoustics.

acoustic niche – The acoustic space, both frequency-based and temporal, occupied by a species in a given environment. By occupying an acoustic niche, a species can hypothetically reduce energy expenditure on communication and avoid acoustic masking or interference from other sources.

A-format – The raw, unencoded output from an ambisonic microphone.

ambisonics – A full-sphere audio format, capable of capturing and reproducing a three-dimensional sound field (represented in B-format). Conventional mono, stereo and multichannel recording formats may be encoded into ambisonics via studio-based processing; conversely, they may be extrapolated—or decoded—from an existing 3D sound field. Higher orders of ambisonic recording and reproduction offer more spatial resolution. See also: A-format, B-format, decoding, encoding.

atmosphere track – A DAW track (or group of tracks) used in audio post-production to add environmental context and ambience to a scene.

autonomous recorder – Any sound recording device capable of operating independently of the user, via a set of pre-programmed instructions. Autonomous recording devices are frequently deployed for extended periods in the field, and are commonly used for passive acoustic monitoring (PAM) in scientific surveys and ecological assessments.

B

baffle – An absorptive object placed between two or more microphones, to enhance acoustical separation between capsules.

bat detector – A device (typically handheld) which allows humans to listen to ultrasonic bat vocalisations.

B-format – A loudspeaker-independent audio format representing a three-dimensional sound field, used in ambisonics.

binaural recording – A three-dimensional stereo recording technique, involving the placement of miniature microphones inside the ears of an artificial or real human head. Intended for playback over headphones, binaural recordings attempt to replicate the real-world perceptual experience of listening, taking account of the human auditory system and the natural filtering effects of our heads and upper bodies on sound. Binaural characteristics may also be recreated virtually, in DAWs and other software.

bioacoustics – The scientific study of animal sounds.

boom pole – A length-adjustable pole, normally made of lightweight carbon fibre or metal, used to position a microphone closer to a sound source.

C

cable-borne vibrations – Vibrations transmitted through the physical structure of an audio cable to a microphone (e.g. by user handling, or natural motion of a mic-mounting surface such as a branch). Generally considered undesirable in wildlife sound recording.

capsule – The transducer portion of a microphone, incorporating the diaphragm.

cardioid microphone – A type of unidirectional microphone with a cardioid (heart-shaped) polar pattern, which is most sensitive to sounds in front of the recording/listening position. Attenuates sounds from the sides and strongly attenuates sounds from the rear.

clipping – In digital audio, clipping occurs when a signal exceeds the maximum possible amplitude range a recording device is capable of representing. In such cases, the upper/lower portions of the signal are flattened (or 'clipped') at the points where the maximum range is exceeded.

closed-back headphones – Headphones featuring a sealed exterior portion on each ear cup, as opposed to an open-back grille. Closed-back headphones provide more passive acoustic attenuation (i.e. non-electronic reduction of sound intensity) than open-backed models, reducing the audibility of potentially distracting environmental sounds during field monitoring. Leakage of monitored sounds from the ear cups, which may be picked up by nearby microphones, is also greatly reduced in closed-back designs.

close-miking – The technique of positioning a microphone as close as is reasonable to a subject, to reveal proximate detail and/or increase signal-to-noise ratio.

contact microphone – A piezoelectric microphone, intended for placement directly onto solid surfaces to transduce internal structural vibrations.

D

decoding – The software-based conversion of an ambisonic sound field into a format suitable for playback over a particular loudspeaker set-up. Decoding can also be used to derive 'virtual' microphone polar patterns, and to construct recording configurations (e.g. XY, mid-side) out of these patterns.

Digital Audio Workstation (DAW) – A graphical software application for recording, manipulating and outputting audio.

double mid-side (DMS) – A three-channel extension of the mid-side recording configuration, in which an additional (rear-facing) mid mic is incorporated to create a forwards- and backwards-facing mid-side array. DMS recording allows for full surround sound capture on the horizontal plane.

drop-rig – An informal term for any recording set-up left unattended in the field, often for extended periods and without the recordist being physically present to monitor signals or make real-time observations. See also: unattended recording, passive acoustic monitoring (PAM).

dynamic microphone – In general discussion (omitting ribbon microphones), refers to a moving-coil microphone which uses electromagnetism to transduce sound into an electrical signal.

E

electroacoustic music – A genre of music emphasising the manipulation and processing of acoustic sound sources in both studio and live-performance contexts, often with an emphasis on multichannel composition and associated performative practices.

encoding – The software-based conversion of sound signals for positioning within an ambisonic sound field.

F

fieldcraft – The accumulated practical knowledge and skill-set required to make successful recordings in the field. Fieldcraft underpins the effective utilisation of audio technologies and is mostly derived from practical, first-hand experience. Examples of wildlife sound recording fieldcraft include: detailed observation and acquired understanding of species behaviour; intuitive understanding of landscape features and their natural effects on sound propagation; knowledge of weather patterns; and the ability to effectively camouflage recording set-ups based on the appearance of native flora.

field recording – Any sound recording activity which takes place in the wider world, outside a purpose-built studio setting.

focus – The globular region within a parabolic reflector, where sound waves running parallel to the reflector's axis of symmetry converge (and where microphone capsules are positioned, in conventional use).

G

gain – The amount of electronic amplification applied to a transduced signal.

H

hydrophone – A fully submersible microphone capable of recording underwater sounds.

L

lapel microphone – Another name for lavalier microphone.

lavalier microphone – A miniature microphone commonly used in production and broadcast applications, affixed to upper-body clothing such as ties, collars and lapels. In wildlife sound recording, lavalier mics are frequently used for drop-rigs, or for placement on structures such as nests, fences and branches during close-miking. Often shortened to 'lav mic'.

leader tape – Non-magnetic tape, historically used by wildlife sound recordists for splicing, marking, adding silent portions to and protecting magnetic tape recordings.

M

maximum sound pressure level (max SPL) – Refers to the loudest sound level a microphone can handle before it begins to noticeably distort.

metadata – Information which describes and/or provides additional context on a field recording. Examples include: location info; weather data; behavioural observations; and notes on microphone configuration. Note that slated (i.e. spoken, field-based) metadata is related to, but distinct from, technical metadata embedded within an audio file.

mid-side (MS) – A coincident stereo configuration in which a figure-eight (bidirectional) microphone is used to capture sounds to the left and right of the recording position, and another microphone (often cardioid) is used to capture sounds in front of the recording position.

monophonic (mono) – Using or referring to a single channel of sound. Compare with: stereo, multichannel.

multichannel – Using or referring to three or more channels of sound. Compare with: mono, stereo.

N

nature sound recording – A term which is largely synonymous with wildlife sound recording. For some, 'nature' may imply a broader approach to audible subjects, with less emphasis on the isolation of individual species and their behaviours from surrounding habitats (using e.g. parabolic reflectors or close-miking techniques).

O

omnidirectional (omni) microphone – A microphone with an omnidirectional polar pattern which, in principle, picks up sound from all directions with equal sensitivity. In practice, omnidirectional microphones become more directional at higher frequencies. Smaller capsules provide a more consistent omnidirectional polar pattern across a given frequency range.

on-axis sound – Any sound originating from a point directly in front of a microphone.

order – See: ambisonics.

ORTF – A near-coincident stereo configuration in which two cardioid microphones are spaced 17 cm apart and positioned at a mutual angle of 110 degrees. ORTF is an initialism of Office de Radiodiffusion Télévision Française, where the configuration was originally developed.

ORTF-3D – A multichannel configuration designed for three-dimensional sound capture, consisting of an array of eight supercardioid microphones.

P

parabolic reflector – A curved surface designed to reflect sound waves onto a focus (within which a microphone capsule, or microphone capsules, may be positioned). Parabolic reflectors enable long-distance species recording with minimal disturbance. They also provide acoustical amplification by design, reducing the need for electronic gain and increasing the overall signal-to-noise ratio of a recording. Also known informally as a 'dish', 'reflector' or 'parabolic mic'.

passive acoustic monitoring (PAM) – Refers to unattended acoustic monitoring of wildlife and ecosystems, typically over a medium- or long-term period and often for the purposes of scientific data-gathering. PAM recording devices may incorporate extensive weatherproofing and other modifications for extended deployments in the field. They may also be programmed to activate at set times, or trigger based on value thresholds set by the user. See also: unattended recording, drop-rig.

polar pattern – A graphical representation of a microphone's directionality, describing its sensitivity to sound arriving from different directions.

post-production – The final stage of audio production in film, TV and related media, incorporating processes such as editing, sound design, mixing, mastering and Automated Dialogue Replacement (ADR).

pre-record buffer – A function on a portable recorder which allows audio to be captured several seconds before the record button is actually pressed, reducing the likelihood of missed opportunities in the field.

Q

quadraphonic (quad) – In a field recording context, using or referring to a four-channel configuration where microphones are positioned in a square arrangement on the horizontal plane.

R

reverberation (reverb) – The continuous blending of multiple sound reflections after an initial sound is produced.

rig – An informal term referring to any recording set-up used in the field.

S

scrim net – A military-type fabric net used to conceal microphones, portable recorders and other audio equipment. Scrim nets are favoured by many recordists for camouflaging microphone windshields because of their relative acoustic transparency.

shock-mount – A microphone support designed to reduce structure-borne vibrations (e.g. from user manipulation of a windshield handle or boom pole, or from the natural movement of a supporting object such as a tree branch).

shotgun microphone – A highly directional tube-shaped microphone, which strongly rejects off-axis sounds via a series of cancellation slots cut into the side of the microphone structure. Despite their high level of directionality, shotgun mics do not provide additional amplification for on-axis sounds.

slate – A verbal announcement containing metadata relating to a recording situation or subject, often made in the field immediately before/after a sound has been recorded.

song-post – A favoured avian vocalising spot, typically used for display or advertising purposes. Examples include prominent branches, telegraph poles and chimney pots.

sound design – The process of creating and shaping sounds for use in productions such as films, TV series and video games.

sound-map – A map-based representation of sounds in an area, often made available online and featuring interactive location markers with embedded recordings.

soundscape – All sounds present in a particular real-world or abstract environment. The concept of a soundscape places emphasis on how sounds are interpreted and perceived by individuals or societies, and the relationships between humans and their sonic environment (as studied through acoustic ecology).

soundscape ecology – Another term for acoustic ecology.

spatial audio – A general term referring to the use of multichannel techniques and 'immersive' audio technologies (e.g. ambisonics) to record and render subjects three-dimensionally.

spectrogram – A frequency-based visual representation of a recording, showing spectral content over time and typically using colour variations to indicate intensity. Often used in combination with a displayed waveform for analysis and editing purposes.

stereophonic (stereo) – Using or referring to two channels of sound. Compare with: mono, multichannel.

Stereo Ambient Sampling System (SASS) – A patented stereo microphone design formerly manufactured by Crown International (USA), featuring two boundary-mounted microphones separated by a specially contoured baffle. In general field recording discourse, this mono-compatible, near-coincident configuration is referred to as 'SASS' or 'SASS-type'. The design is widely emulated in custom mic-building projects.

T

tree-ears – Any recording technique involving the attachment of microphones to a tree, typically using the trunk or a large branch as a mounting surface.

U

ultrasound – Any audio frequency above the upper limit of human hearing (c.20 kHz).

unattended recording – Any recording made without the recordist being immediately present in the field. See also: drop-rig, passive acoustic monitoring (PAM).

V

vocabulary recording – A wildlife sound recording methodology which involves recording a species' vocal repertoire as fully as possible, often over a long-term period and with an emphasis on documenting previously unknown vocalisations (and audible behaviours) for the purposes of bioacoustics research.

W

waveform – An amplitude-based visual representation of a recording. Often used in combination with a spectrogram for analysis and editing purposes.

wildlife sound recording – A specialised form of field recording, focusing on sounds produced by undomesticated species behaving freely within their associated habitats. See also: nature sound recording.

wind noise – In a technical context, refers to the sound of wind acting across the physical structure of a microphone or microphone windshield. Generally considered undesirable in wildlife sound recording.

windshield – A structure placed around a microphone, to shield against the unwanted effects of wind and provide general protection. Very broadly, there are two main types of windshield: fully enclosing baskets (also known as 'blimps' or 'cages') and slip-on covers. Both types may incorporate, or allow the addition of, faux-fur or fleece-based covers for additional protection.

X

XY – A coincident stereo configuration in which two unidirectional or bidirectional microphone capsules are positioned as closely together as possible, at a specified mutual angle. The configuration renders spatial information via level differences between signals arriving at each capsule.

32-bit float recording – A recording format capable of representing a huge dynamic range of over 1,500 dB, which removes both the possibility of digital clipping and, in many field situations, the need for input gain adjustments on portable recorders.

Index

References to figures appear in *italic* type.